T0313206

Image-Based Damage Assessment for Underwater Inspections

Image-Based Damage Assessment for Underwater Inspections

Michael O'Byrne
Bidisha Ghosh
Franck Schoefs
Vikram Pakrashi

CRC Press
Taylor & Francis Group
Boca Raton London New York

CRC Press is an imprint of the
Taylor & Francis Group, an **informa** business

CRC Press
Taylor & Francis Group
6000 Broken Sound Parkway NW, Suite 300
Boca Raton, FL 33487-2742

© 2019 by Taylor & Francis Group, LLC
CRC Press is an imprint of Taylor & Francis Group, an Informa business

No claim to original U.S. Government works

Printed on acid-free paper

International Standard Book Number-13: 978-1-138-03186-9 (Hardback)

Library of Congress Cataloging-in-Publication Data

Names: O'Byrne, Michael (Structural engineer), author. | Ghosh, Bidisha, author. | Schoefs, Franck, author. | Pakrashi, Vikram, author.
Title: Image-based damage assessment for underwater inspections / Michael O'Byrne, Bidisha Ghosh, Franck Schoefs and Vikram Pakrashi.
Description: Boca Raton : Taylor & Francis, a CRC title, part of the Taylor & Francis imprint, a member of the Taylor & Francis Group, the academic division of T&F Informa, plc, [2018] | Includes bibliographical references and index.
Identifiers: LCCN 2018005237 (print) | LCCN 2018029841 (ebook) | ISBN 9781351052573 (Adobe PDF) | ISBN 9781351052566 (ePub) | ISBN 9781351052559 (Mobipocket) | ISBN 9781138031869 (hardback) | ISBN 9781351052580 (ebook)
Subjects: LCSH: Offshore structures--Inspection. | Underwater imaging systems.
Classification: LCC TC180 (ebook) | LCC TC180 .G37 2018 (print) | DDC 627/.7020287--dc23
LC record available at https://lccn.loc.gov/2018005237

Visit the Taylor & Francis Web site at
http://www.taylorandfrancis.com

and the CRC Press Web site at
http://www.crcpress.com

Contents

4 Fundamentals of image analysis and interpretation   49

5 Crack detection   79

6 Surface damage detection   97

Preface

Marine structures play an important role in the economic and social activities of the regions in which they serve. Many residential, commercial, industrial, and transport infrastructures are located along our coastlines, while structures located far offshore supply a growing portion of our energy needs. These structures must typically operate in harsh and corrosive conditions, and for this reason, they are particularly susceptible to rapid aging and deterioration. Inspections are therefore crucial to ensure that safety and structural integrity are maintained at an acceptable level.

Historically, the monitoring of marine structures has been based on recurrent visual observations and assessments of structural condition carried out by trained divers. Recent research efforts have centered on developing effective and reliable methods for acquiring, managing, integrating, and interpreting structural performance data at a minimum cost while reducing the unreliable human element. In this vein, image-processing in underwater inspections has garnered a lot of attention as a way of strengthening existing visual inspection methods by introducing a source of quantitative information that naturally complements the largely qualitative information obtained from conventional visual inspections.

Image-processing in underwater inspections is a relatively new field and has developed considerably over the past few years. It owes its rising popularity to several factors: data collection via optical sensing is an inherently quick, clean, inexpensive, and versatile non-contacting process, and unlike other sensing methods, cameras require minimal training in their operation, can be easily adapted for underwater application, and the acquired data is easy to visualize and interpret. Moreover, vision (both human and machine) is often the only way to sense anomalies such as cracks or surface corrosion. Image-processing algorithms can be used to automatically detect these types of anomalies in inspection imagery and extract quantitative information such as the size and shape of defects. While many image-processing methods have been devised for topside/terrestrial structural health monitoring applications, such as damage detection, vibration assessment, and 3D shape recovery, these methods often encounter performance issues when applied to underwater imagery because of the reduced visibility.

Prior to this book, little attention has been given to devising new algorithms, or modifying existing ones, for dealing with the hurdles imposed by the challenging underwater environment.

This book takes a holistic view of imaging in underwater inspections and touches on the central questions that will confront new researchers and practitioners in this field. In order for inspectors to properly capitalize on the power of cameras and integrate them into the inspection framework, they should first be aware of the best practices when capturing imagery on-site. A set of guidelines for obtaining imagery such that the acquired imagery is well-suited for subsequent quantitative analysis is outlined. The fundamentals of image processing are covered, and subsequently, some of the most relevant algorithms for inspectors are described in a step-by-step manner. These algorithms include crack detection, surface damage detection (e.g., for detecting corrosion), and stereo-based 3D imaging. Readers can run these algorithms themselves using the MATLAB®* scripts that are included with this book, as well as benefiting from access to a large collection of underwater images.

It is important for inspectors to know the effectiveness of techniques when applied in the field. For this reason, a methodology for evaluating the performance of image-processing techniques under various underwater visibility conditions is presented. Finally, we look at how the information obtained from image analysis can be used in the broader context of SHM.

The material in this book is explained with the help of real-world case studies gleaned from the authors' experience of working in this field over the last decade. Overall, it is hoped that this book can act as a helpful resource for structural engineers, inspectors, and researchers in this field, who are looking to effectively integrate imaging into the inspection regime.

*For product information, please contact:

The MathWorks, Inc.
3 Apple Hill Drive
Natick, MA 01760-2098 USA
Tel: 508-647-7000
Fax: 508-647-7001
E-mail: info@mathworks.com
Web: www.mathworks.com

Authors

Dr. Michael O'Byrne (Ph.D.) is a post-doctoral researcher in the School of Engineering at University College Cork, Ireland. His research interests are image-based non-destructive testing techniques for monitoring offshore structures. His doctorate investigated Automatic Detection of Damage using Image-Based Techniques in Underwater Marine Structures. It involved using the latest research in image processing to detect and quantify damage that affects offshore structures, such as cracks, corrosion, and bio-fouling. Currently, Dr. O'Byrne is developing new image processing techniques for infrastructure maintenance management and also looking at how underwater image processing based inspection can help estimating changes in hydrodynamic loads to structures due to bio-fouling.

Dr. Bidisha Ghosh (Ph.D.), assistant professor, Trinity College Dublin, is an expert of statistical modelling, artificial intelligence techniques, and data analysis. She applies these techniques to transportation networks, hydrological networks, and infrastructure management. Her work in image processing relates to structural damage detection, infrastructure management, traffic monitoring, crash-barrier design, and the development of a benchmark repository for such purposes.

Professor Franck Schoefs (Ph.D.), from the University of Nantes, France, is a leading figure in the field of structural reliability and inspection-led maintenance management. He works on probabilistic modelling of inspections results and on site measurements from structural health monitoring. Major applications of his work are in bridge engineering, offshore structures, and marine renewable energy. He is an expert of probabilistic modelling of marine growth on offshore structures.

Dr. Vikram Pakrashi (Ph.D.) is assistant professor and director of Dynamical Systems and Risk Laboratory. His research interests strongly feature infrastructure maintenance management and Structural Health Monitoring. Dr. Pakrashi has experience of inspecting, instrumenting, and assessing numerous damaged structures at different levels of complexity and detail.

Chapter 1

Introduction

Offshore and marine engineering have developed rapidly in recent decades. Advances in construction materials and methods have paved the way for new forms of structures, such as wave energy converters, to be installed in deeper and more hostile conditions. Inevitably, however, newly installed structures will join the existing infrastructure stock, and the focus will shift toward monitoring and maintenance activities to ensure that these structures reach, or even surpass, their target lifespans (Wenzel, 2003).

The harsh and unrelenting marine environment means that offshore and marine structures are prone to aesthetic, functional, or structural degradation, which, over time, typically leads to a loss of serviceability at either a component or global level. Owners/managers must, therefore, inspect structures to ensure that they are safe, fit for service, and so that they can make more informed decisions when allocating resources toward the correction of deficiencies. This latter aspect has attracted a growing interest in recent times as the importance of life cycle optimization and related financial benefits continue to be recognized, especially in relation to marine structures (Schoefs et al., 2012). Given that many significant decisions are made based on inspection findings, it is important that inspections strive to provide accurate information that is reflective of the true condition of the structure.

This book highlights the value that image-processing can bring to the inspection process, especially in relation to, but not confined to, underwater inspections. The versatility of image-processing is conveyed through the wide range of applications presented in this book.

1.1 AIM OF THIS BOOK

The goal of this book is to serve as a comprehensive guide for structural engineers and inspectors on how to effectively integrate image-processing techniques into the inspection regime. Image-based inspections are comprised of several distinct, yet inexorably linked, stages, as illustrated in Figure 1.1. The success at each stage has a significant bearing on the overall

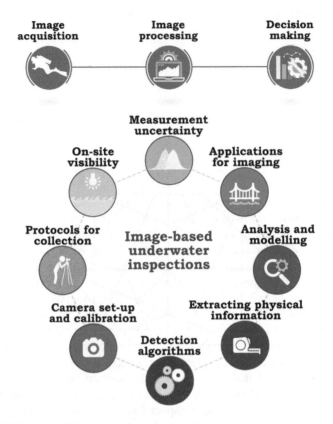

Figure 1.1 Elements of image-based underwater inspections.

success; failure to ensure that each stage is given due consideration will almost invariably result in a reduced quality of the inspection results. In this book, a holistic view of image-based inspections is adopted, and each stage is comprehensively addressed.

The three principal stages of imaging in underwater inspections are covered. These are:

1. On-site image acquisition procedures
2. Image-processing algorithms
3. Using the results from image-processing for further analysis with specialized engineering software

A firm grasp of these stages provides the building blocks necessary for readers to become conversant in using imaging in inspections. The first pillar sets out best practice guidelines for obtaining imagery that is well-suited for subsequent quantitative analysis.

For the second stage, a concise and relevant overview of the fundamentals of image-processing is presented. Following this, we delve into more advanced damage assessment algorithms. These algorithms are described in a step-by-step manner as the goal here is to give readers a thorough understanding of the algorithmic design choices and to equip readers with the skills necessary to modify algorithms for their purposes. MATLAB® scripts are provided, and readers have access to a large underwater image repository so that techniques can be run and tested on a collection of realistic image samples.

For the third and final stage, we look at how the information obtained from image analysis can be used in the broader context of Structural Health Monitoring (SHM). The material in this book is explained with the help of real-world case studies. It is hoped that these studies will give readers a glimpse of the potential application areas of image-processing.

This book is principally geared toward the inspection of offshore and marine structures; however, many of the concepts and techniques presented in this book can be applied to terrestrial/top-side inspections.

1.2 IMAGING IN INSPECTIONS

Image-processing in underwater inspections is still a young field. Work supporting the development and implementation of image-processing methods has only begun to emerge relatively recently, and its position within the inspection framework is not well defined. The role of cameras is typically limited to taking photographs or video of instances of damage without any agreed protocol of image collection and subsequent interpretation within the inspection framework (Phares et al., 2004). The collected imagery is usually archived and is rarely ever used in a quantitative sense. However, there exists significant scope for further development in the domain of underwater SHM as, currently, the full potential of cameras is not being realized.

Image-processing is well-suited to complementing traditional visual inspection methods as opposed to completely replacing them. Visual inspections are the most common means of collecting data about the state of marine structures; however, they have some inherent limitations. They are affected by the ability of inspectors to observe and objectively record details of defects, and they are prone to considerations such as boredom, lapses in concentration, subjectivity, and fatigue, all of which contribute to greater variability and reduced accuracy (Dirksen et al., 2013; Estes & Frangopol, 2003). Image-processing provides a way of overcoming some of these shortcomings by making visual data a part of quantitative assessment.

1.2.1 Advantages of image-processing as an inspection tool

Assessing the submerged part of marine structures introduces new challenges for inspectors. Many damage diagnostic tools that can be used on dry land cannot be readily adapted for underwater deployment. Additionally, only a limited amount of time can be spent underwater, especially when the inspection is being carried out by a diver rather than a remotely operated vehicle (ROV). This puts an emphasis on adopting expeditious data collection practices. Unlike other non-destructive testing (NDT) tools, cameras can acquire data in an efficient manner, they are easily adapted for underwater application, and they require only minimal training in their operation. Moreover, vision-based NDT tools are often the only practical way to detect certain damage indicators such as cracks, corrosion, or surface breaking defects.

Cameras are already used in almost all visual inspections. Typically, photographs are captured to include in the inspection report to accompany the inspector's comments; however, the photographs are rarely exploited to their fullest potential in either a qualitative or a quantitative fashion. Adopting effective image-based techniques can provide accurate quantitative information with minimal human supervision to supplement visual inspection techniques and increase reliability. The quantitative nature of the data obtained from image analysis naturally lends itself to numerous applications—it is helpful for developing new damage models or strengthening existing ones, which are used to forecast the rate of propagation of damage as the structure continues to operate.

1.2.2 Limitations

While image-based methods undoubtedly have the potential to be a convenient and useful underwater NDT tool, there are some limitations as well. First, the technology is only appropriate for assessing visible damage forms and surface breaking defects. It is not possible to assess internal defects or damage that is masked by bio-fouling without first cleaning the structure. Second, the poor underwater visibility conditions diminish the ability of cameras, and subsequent image-processing techniques, to effectively identify instances of damage. Underwater images are affected by light attenuation, scattering, color absorption, suspended particles, and air bubbles, as identified by Massot-Campos and Oliver-Codina (2015), who carried out a survey on optical sensors and methods for 3D reconstruction in underwater environments. These issues lead to blurriness, reduced contrast, and an overall loss of image quality. This problem is exacerbated in high turbidity conditions or when intense artificial light sources are employed. Artificial light sources often create highly non-uniform lighting in the scene, causing specular reflections that mask detail and create bright spots that may mislead damage detection algorithms. This book addresses this problem by presenting a repository driven approach for evaluating the performance of methods under various visibility conditions.

Finally, there is still a gap between image-processing algorithms and humans when it comes to semantic scene interpretation. Image-processing techniques can struggle with tasks that are intuitive to humans such as recognizing instances of damage. As such, there will be many situations where inspectors cannot fully rely on image-processing techniques without some level of supervision or oversight. Nevertheless, the field of computer vision is continually advancing, and emerging algorithms show promising signs that should help bridge this gap. There is an ongoing proliferation of new and improving image acquisition devices and technologies that will help shape the future of image-based underwater inspections. For now, knowing which situations stand to gain from using image-processing, and knowing how to develop algorithms that are most effective in these situations, are key ingredients in realizing the full power of cameras and image-processing algorithms.

1.3 SAMPLE APPLICATIONS OF IMAGE-PROCESSING TECHNIQUES

The following sample applications represent some of the author's experiences in structural monitoring and span the range of motivation for using image-processing techniques. Like many of the examples presented throughout this book, these case studies highlight the relevance and limitations of the latest image-based technology.

1.3.1 Measuring the width and length of cracks

Assessing the extent of cracks is often a key part of underwater inspections. The presence of cracks provides an indication of the structural degradation and is an important factor when diagnosing the condition of a structure. For the particular case of offshore structures, the structural deterioration phenomenon is mainly due to fatigue caused by waves acting continuously on the steel elements. The decrease of the resistance of the structural capacity over a time interval is caused by the cracking of the tubular joints. The inspection of an offshore structural system has two objectives: (a) to detect both the presence and the size increase of cracks and (b) to perform the necessary maintenance actions to the structure accordingly. For marine structures in general, insight into the crack growth development is crucial for guaranteeing structural reliability and for scheduling efficient maintenance schemes.

Automatic crack detection techniques are of particular interest to engineers/ inspectors as they remove the need for inspectors to manually count and

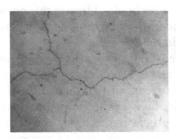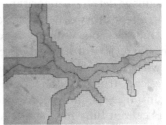

Figure 1.2 **Image-based crack detection.**

measure cracks, which is often a dull, subjective, and error-prone task. This problem is magnified when performed in challenging subsea conditions. Given the pervasive nature of cracks and the value that crack detection techniques can bring to the inspection process in terms of increased efficiency and convenience, it is worthwhile to devise performant methods. In Chapter 5, we explore methods for detecting cracks, as depicted in Figure 1.2, and subsequently quantifying certain properties like the crack length and crack width.

1.3.2 Automatic corrosion detection

The motivation behind the development of image-based damage detection techniques is to provide inspectors with an efficient way to locate instances of damage and rank their severity (i.e., based on size). This is helpful when prioritizing parts of a structure for repair work or electing parts of a structure to be scrutinized further using more localized NDT tools, such as ultrasonic imaging, which can give a sense of the extent of damage below the surface.

Most damage detection algorithms consist of image segmentation followed by subsequent classification of the segmented regions. Two examples of image-based damage detection are shown in Figure 1.3, where corrosion is detected on the surface of marine structures. Segmentation algorithms use either color information, texture information, or a combination of both to isolate similar regions in an image. Naturally, the effectiveness of color based segmentation algorithms and texture based segmentation algorithms

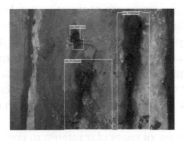

Figure 1.3 **Image-based corrosion detection.**

will vary according to the surface and damage type under consideration as certain damages are more separable from the undamaged surface based on either their color or texture attributes.

The main forms of surface damage encountered on aging infrastructural elements (corrosion, leaching, etc.) are often characterized to a greater extent by the change in color from the undamaged surface than a change in texture, while conversely, texture based segmentation is well-suited for specific applications such as spalling of concrete where the damaged regions have similar color yet rougher texture than the surroundings. With this in mind, this book looks at methods for segmenting damage based on color information as well as considering texture based segmentation methods.

1.3.3 Bridge vibration assessment

Image-processing techniques can be readily extended for video analysis, which opens the possibility of studying problems of a dynamic nature. Video analysis has the potential to be used as a substitute for many instrumentation devices, such as the complicated networks of accelerometers that are often used to monitor the health of marine structures. One specific application in SHM where this is of relevance concerns characterizing the dynamic response of vibrating structures.

In Chapter 9, we describe a video tracking technique and the camera set-up for assessing the time-varying bridge displacements for the purpose of identifying the natural frequency. The video tracking technique operates by tracking the movement of a point on the bridge while in an excited state. A plot of the recorded bridge vibrations is illustrated in Figure 1.4. The only equipment required is a conventional digital video camera.

1.3.4 3D shape recovery of marine growth colonized structure

The inspection of marine growth is a small, but significant, part of the whole annual survey program of offshore structures. The formation of marine growth on offshore structures introduces several problems; most notably, it increases drag forces and creates unpredictable hydrodynamic effects. These factors often cause a loss in structural performance and reliability, which result in a shortened lifespan for the structure. Engineers, therefore, need to obtain accurate data to assess the overall extent of marine growth, the relative proportions of hard and soft growth, the average thickness and percentage cover of the marine growth layer, and the weight of growth, particularly for extensive hard marine species. Ideally, a complete inspection of biofouling on a structure should be designed to periodically collect the following information: (a) identification of marine growth species constituting the different level of biofouling, (b) thickness of superimposed

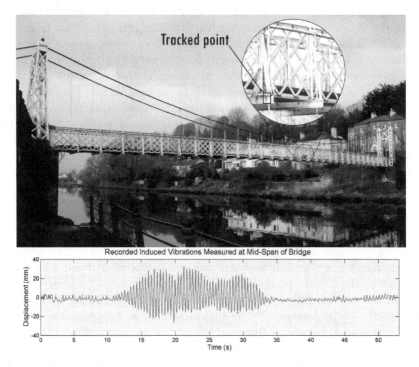

Figure 1.4 Video tracking for assessing the dynamic nature of a vibrating bridge.

layers and their weight, (c) percentage of surface covered by the main species, and (d) the average size of each species present on the structures.

While there has been extensive research carried out on the damaging effects of marine growth in general, there has been little work done on developing accurate models that reflect the complex shapes and roughness properties of marine growth that is encountered in reality. Improved knowledge of the marine growth measurements can produce more accurate models for structural reliability and risk assessment. In Chapter 7, we describe an underwater 3D imaging system and technique for recovering the full 3D shape of submerged structural elements, as depicted in Figure 1.5 for a monopile structure.

The value of accurate 3D models for subsequent analysis tasks is substantial. One high valued workflow involves feeding the 3D model of marine growth affected structural member into a finite element (FE) software package. From this, the forces under certain known wave conditions can be analyzed. This workflow is discussed in Chapter 9.

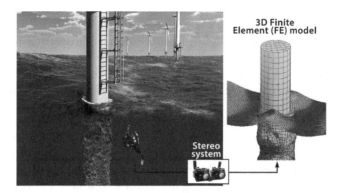

Figure 1.5 Using stereo imaging to recover 3D shape of marine growth colonized structure.

1.4 BOOK OUTLINE

This book is divided into four main sections. The first section lays out the context for imaging in underwater inspections and describes the background and current status of inspections. The second section covers the fundamentals of image-processing and discusses the image acquisition procedures. Here, we guide inspectors/divers through the stages of image acquisition, and we place an emphasis on adopting systematic approaches for on-site image collection so that the acquired imagery is well-suited for quantitative analysis. Following on from this, the third section describes the algorithms and methodologies for assessing damage. The techniques strategically include a crack detection technique, two surface damage detection techniques (one based on color information and one based on texture information), and a 3D imaging technique using a stereo based system. The crack detection and surface damage detection algorithms are the building blocks for most damage detection tasks, while 3D imaging is useful as it can be used for volumetric measurements, documentation, and presentation to the general public. We adopt a hands-on approach in which we give a step-by-step breakdown of how our solutions are designed and deployed, and we offer practical insight into why certain algorithmic design choices are made.

Finally, the fourth section considers the future applications in this emerging field. Promising technologies that will likely shape underwater inspections in the years to come are investigated, and in addition, the improvements that can be made to current methodologies and workflows are discussed.

1.5 SUMMARY

This book provides an ideal introduction to image-processing techniques for structural engineers and inspectors who wish to harness the full potential of cameras as an inspection tool. Improving the quality of inspection data has become an area of growing interest as the optimal management of infrastructure assets and their continued integrity is increasingly valued by owners/stakeholders. This book aims to enhance the quality of underwater inspections through the use of image-processing based techniques, which is a very important but not very developed field of study as of yet. This research advances the field in three ways. We look at image acquisitions procedures, image-processing algorithms, and how the results from these algorithms can be used by engineers or decision makers. Considered collectively, these three areas provide a comprehensive and well-rounded guide for readers looking to use imaging as part of the inspection regime. This book can be expected to be used by inspectors, owners, and managers of marine engineering structures. Although the developed techniques are focused on underwater application, they are quite general and can be readily used in a much broader context.

REFERENCES

Dirksen, J., F. H. L. R. Clemens, H. Korving, Frédéric Cherqui, Pascal Le Gauffre, T. Ertl, H. Plihal, K. Müller, and C. T. M. Snaterse. "The consistency of visual sewer inspection data." *Structure and Infrastructure Engineering 9*, no. 3 (2013): 214–228.

Estes, Allen C., and Dan M. Frangopol. "Updating bridge reliability based on bridge management systems visual inspection results." *Journal of Bridge Engineering 8*, no. 6 (2003): 374–382.

Massot-Campos, Miquel, and Gabriel Oliver-Codina. "Optical sensors and methods for underwater 3D reconstruction." *Sensors 15*, no. 12 (2015): 31525–31557.

Phares, Brent M., Glenn A. Washer, Dennis D. Rolander, Benjamin A. Graybeal, and Mark Moore. "Routine highway bridge inspection condition documentation accuracy and reliability." *Journal of Bridge Engineering 9*, no. 4 (2004): 403–413.

Schoefs, F., Boéro, J., Clément, A., and Capra, B. "The αδ method for modelling expert judgment and combination of NDT tools in RBI context: Application to marine structures, structure and infrastructure engineering: Maintenance, management, life-cycle design and performance (NSIE)." *SMonitoring, Modeling and Assessment of Structural Deterioration in Marine Environments 8*(Special Issue), (2012): 531–543.

Wenzel, H. "E-MOI-European monitoring initiative." *Proc. SHMII-1, Structural Health Monitoring and Intelligent Infrastructures 1* (2003): 145–152.

Chapter 2

Inspection methods and image analysis

2.1 INTRODUCTION

Inspections are essential for ensuring that marine structures remain safe and fit-for-service. This chapter discusses the recent history of underwater inspections and provides an overview of the various testing tools that inspectors rely on to detect and diagnose damage. The typical roles and duties of inspectors are outlined, and ways in which image-processing techniques can strengthen conventional inspections are explored.

Image-processing techniques have wide applicability in the domain of underwater inspections. They can identify many common damage forms or features of interest encountered in marine structures and record in-situ measurements from underwater scenes. We look at the benefits of utilizing image-processing based methods as part of the inspection regime from an on-site implementation standpoint, as well as from the point of view of how image-processing methods serve to enhance the overall quality of information gathered from underwater inspection campaigns. Some of the limitations of image-processing based methods, especially in comparison to other non-destructive testing (NDT) methods, are also addressed.

Future developments and opportunities on integrating image-based methods into the inspection framework more effectively are highlighted. This chapter concludes with a discussion of how image-based methods can be more effectively and systematically integrated into the inspection framework.

2.2 INSPECTION OF MARINE STRUCTURES

Marine structures are particularly susceptible to rapid aging and deterioration due to the harsh operating conditions (Stacey et al., 2008). Deterioration in marine environment commonly occurs via cracking, corrosion, leaching, biological attach from marine organisms, and other chemical attacks (Chandler, 2014). The effect of such deterioration can be aesthetic, functional, or structural in nature, and in most cases, the result is a loss of serviceability at either a component level or at a global level.

Depending on the jurisdiction, inspection of marine structures at regular intervals may even be mandated by law (Moan, 2005). Nevertheless, owners/operators of structures are incentivized to carry out inspections for financial and safety reasons. Effective monitoring is essential for identifying damage, and early intervention prevents more serious problems from arising down the line that would require more costly and extensive repair work or other intervention measures. In-service monitoring also plays a key role when it comes to deciding whether to extend the lifespan of a structure. Marine structures are typically designed for predefined periods of between 20 to 50 years, depending on their function (Boéro et al., 2012). After this period, most well-maintained structures are eligible for requalification schemes that prolong their service life. In many cases, keeping the structure operational for longer is a more financially attractive alternative to destruction (Rouhan & Schoefs, 2003). On the other hand, structures that are not well-maintained inevitably succumb to the harsh marine environment (Gudmestad, 2000). Once these structures are rendered severely structurally deficient or functionally obsolete, the exorbitant restoration costs are highly prohibitive of lifespan extension plans.

Despite the significant financial rewards associated with effective monitoring, owners of infrastructure are often limited or unwilling to allocate money toward the process of monitoring itself. According to a report by the American Society of Civil Engineers, $22 billion was provided to maintain and build upon the existing inland waterways and marine ports infrastructure network in the United States; however, it is estimated that investment of $37 billion is required (Petrequin, 2011). Given that short-term cost-effectiveness is becoming an unavoidably compelling factor for such decisions, there is a strong emphasis on adopting inexpensive inspection strategies that can still obtain high-quality information about the health of structural components. This is especially pertinent for very large submerged structures or marine structures with a significant number of components where there is a potential for failure. As these inspections are expensive to begin with, an effective cost-saving measure has significant potential impact.

The optimal inspection plan would be to inspect at the right place, at the right time, at the lowest cost, and with the right damage diagnostic tool. While choosing the right time and place to inspect is largely beyond the scope of this book, inspectors looking to choose a low-cost tool that can detect and quantify many of the damage forms encountered on coastal and marine structures (e.g., cracking, corrosion, marine growth colonization) can consider image-processing based methods as a practical option. Visual inspections typically carried out for marine structures indicate that image-processing methods can naturally strengthen conventional visual inspections by introducing a source of quantitative information. Alternative damage diagnostic tools are also available and can complement or enhance camera-based approaches.

2.3 STATUS OF INSPECTION PROCESSES

2.3.1 Types of inspections

Inspection procedures vary depending on the type of structure. For instance, inspection of offshore oil and gas platforms have different requirements to inspection of marine vessels. Comprehensive testing and inspection standards for various marine infrastructure are issued by Det Norske Veritas (DNV) and are available online (Det Norsk Veritas, 2003). Broadly speaking, there are some general inspection types and well-recognized practices that apply across the board for maritime inspections. Inspections can be broadly partitioned into three categories, namely, (i) Routine, (ii) Principal, and (iii) Special inspections. The frequency and exact nature of each inspection type vary somewhat in different jurisdictions. As an example, the inspection and maintenance guidance for port structures in the UK has been drawn from the UK Highways Agency Design Manual for Roads and Bridges (DMRB) BD 21/01 Inspection of Highway Structures (Design Manual for Roads and Bridges, 2001). The reasoning behind this is that the consequences of any failures associated with port structures will be similar to those of a damaged bridge, and the environmental conditions are comparable. Table 2.1 summarizes the categories of inspection contained within the DMRB guidelines.

As a first step for all inspections, some preparation work should be carried out. This includes gathering information about the history of the structure, visiting the structure for an initial appraisal, drawing up an assessment plan, and collating all available structural, environmental, and past service information. Careful planning in this manner yields dividends during the actual inspection and improves the overall efficiency. The following subsections provide a more detailed overview of the main inspection types.

Table 2.1 Categories and frequency of inspection as per the DMRB guidelines

Type of inspection	Description
Routine/General inspection	Visual inspection every 2 years.
Principal inspection	Detailed visual inspection every 6 years.
Special inspection	Inspection after a special event, e.g., a storm.
Safety inspection	Inspection initiated where a defect has been identified.
Inspection for assessment	Inspection to determine load carrying capacity of a structure and its resistance.

Figure 2.1 Photographs are collected as part of routine inspections. The aim of these inspections is to ensure structures are operating as intended and that there are no major issues.

2.3.1.1 Routine inspections

Routine inspections involve a cursory check for gross defects rather than a detailed, close-up examination. They are carried out to ensure that structures are operating as intended and to verify that there are no significant maintenance or safety issues. The outcome of routine inspections may be a recommendation to carry out remedial works or to change the frequency of inspections. Routine inspections are usually carried out on a biannual basis for many marine structures; however, structures in less aggressive environments may be inspected less frequently. Routine inspections do not involve any special NDT tools, but a photographic record is typically compiled as part of the inspection process (Figure 2.1).

2.3.1.2 Principal inspections

Principal inspections involve a visual inspection of all parts of the structure, including inaccessible parts of the structure (Figure 2.2). The purpose of these inspections is to assess the need for repair work, keep track of

Figure 2.2 Principal inspections involve inspecting inaccessible components.

changes to the condition of structural components, and check whether routine maintenance is being properly carried out.

These inspections cover a range of detail, from a close examination of all surfaces to a broader-scale visual inspection. Damage is usually qualitatively described and archived by the inspector. Specialized NDT tools are not required; however, digital photographs are almost always captured to complement the damage descriptions. An attempt to quantify the severity of the damage is generally made using a numerical scale, which typically ranges over a limited number of categories (e.g., 5 levels), leading to a significantly varied degree of uncertainty and vagueness and, consequently, more errors when classifying critical components or structures. Additionally, while numerical scales are helpful for relative ranking and prioritizing, they are not easily integrated into future quantitative analyses or experiments and repair options as this level is usually based on qualitative comments based on personal expertise.

The principal inspection report typically contains all inventory data and all data from the latest principal inspection on one structure. These reports usually contain photographs taken during the inspection to qualitatively convey the condition of components of a structure, or the structure in general. These reports also usually provide a chronological overview of the structure, indicating dates and condition ratings for previous inspections. The outcome of principal inspections may include a cost estimate for repairs that is based on a standard list of component repair rates. If repair work is required, then a recommended or estimated timeframe to carry out the works is also normally provided. There is also a recommendation on when to conduct the next principal inspection. In some cases, the inspector can also call for a follow-up special inspection. A special inspection is warranted if the inspector is uncertain about the extent of damage, is concerned about an existing damage and requires a better quantitative and qualitative understanding of the condition of the materials of the structure, or is unsure about how to proceed with damage repairs based on visual investigations alone. Principal inspection intervals usually range from one to five years depending on the condition and age of the structure, the operating environment, and the expected rate of deterioration. Principal inspections are performed by a qualified person, typically a trained engineer.

2.3.1.3 Special inspections

Special inspections are carried out to determine the nature, extent, and cause of damage to a structure. The inspection necessitates a detailed assessment of the damaged structure. Special inspections involve non-destructive testing with the use of specialized equipment, and possibly destructive testing if deemed necessary. Destructive testing may involve removing material samples for further investigation in a lab environment.

Special inspections are often costly and require extensive effort owing to their in-depth and comprehensive nature. The costs can be subdivided

Figure 2.3 Specialized NDT tools are often used as part of special inspections. Underwater inspections are considered to be special inspections due to the added challenges.

into fixed costs, time-dependent costs, and reliability-dependent costs (Wall et al., 1998). The fixed costs relate to expenses stemming from equipment, site preparation, arranging safe access for inspectors, training, etc. The time-dependent costs accrue over the course of the inspections. This includes any loss of revenue that results from the structure being taken out of operation. As such, expeditious inspections are desired to minimize time-dependent costs. The reliability-dependent costs include the cost of repairs for any defects found, the cost of reinspection or repair of spurious defects, and costs associated with missing defects that may propagate and lead to more severe damage. To achieve a successful and economically viable inspection, inspectors must take all of these costs into account and weigh up the risks and benefits when comparing expeditious and relatively unreliable inspection methods with expensive and reliable methods.

Guidance is provided usually with regards to the most suitable rehabilitation scheme for the structure. Several relevant repair strategies are normally suggested including an option of "do nothing." A cost-benefit analysis of each strategy is undertaken, which includes investigating the direct and indirect costs associated with each option. The indirect costs account for delaying remediation work and thereby allowing the structure to degrade further, as well as any inconvenience that this presents to end users. A standard financial planning formula, such as net present value (NPV), can be employed when determining the optimal strategy. Underwater inspections are often regarded as special inspections given the additional overhead and operational complexities compared with topside inspections (Figure 2.3).

2.3.2 Underwater inspections

Assessing the submerged part of marine structures introduces new challenges for inspectors. These challenges are thoroughly summarized by Busby (1979) and Ramos (1992). These works detail the legal requirements of undertaking underwater inspections, the modes of inspection, and the available damage diagnostic tools. Many damage diagnostic tools that can be used on dry land cannot be easily adapted for underwater deployment. Additionally, only a limited amount of time can be spent underwater, especially when the inspection is being carried out by a diver rather than a remotely operated vehicle (ROV). This puts an emphasis on adopting expeditious data collection practices that harness the full potential of available tools.

Divers and ROVs are the dominant modes for carrying out underwater inspections, although manned submersibles and atmospheric diving suits (ADS) are also options (Figure 2.4). Divers offer significant dexterity and maneuverability. They can access zones that are beyond the reach of larger ROVs, such as in between members of offshore jacket-type platforms. Divers can also carry a variety of NDT tools; however, the use of any specialized equipment requires additional training in their operation. In contrast, ROVs cannot easily be fitted with new NDT tools, and NDT tools that are compatible with an ROV will be more expensive than standard NDT tools of the same type. Trained divers can descend to depths of up to 50 m, but this requires special certification, and it is more common for divers to operate at depths no greater than 30 m.

ROVs may be equipped with numerous sensors. Typically, they carry at least one or more camera/video systems. They can be deployed to much greater depths and for longer periods than divers. They are particularly useful for providing video data for extremely deep operations or for places that are inaccessible or too hazardous for conventional diving, such as polluted, contaminated, or intensely cold water. Although the dependability and functionality of ROVs are steadily increasing, some limitations persist. They are considerably more expensive, inflexible, and vulnerable to failure in comparison to diver-based approach. They still require heavy lifting equipment and specialist support teams, including highly trained pilots who must usually possess formal qualifications. Micro-ROVs are increasingly being used as part of underwater inspections. They are affordable (typically around a few thousand US dollars) and can operate in depths up to 300 m. Micro-ROVs usually weigh in under 10 kg, which means

 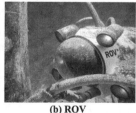 

(a) Diver   (b) ROV   (c) Atmospheric diving suit

Figure 2.4 (a–c) Modes of inspection.

they are highly portable. Their small dimensions also mean that they can access structural components that are beyond the reach of full-size ROVs. Basic micro-ROVs come with a video camera unit and lights, while more advanced ones may be equipped with high-quality multi-camera systems and sonar. Because of their size and thrust capabilities, operators may encounter problems controlling micro-ROVs in rough sea conditions or in the presence of strong currents. Consequently, their safe operation requires great attention to detail.

As things stand, diver-based approaches remain the current state-of-the-art owing to their versatile, economical, and easily deployable nature.

2.3.3 Visual inspections carried out by divers

Visual inspections are the most straightforward form of inspection and are the backbone of most underwater inspections (Watson, 1992). The diver is presented a brief outlining the task at hand and specifying the target sites and certain things to look out for. Depending on the type of structure, divers may be asked to check for the following:

- Corrosion or indicators of corrosion
- Deterioration of the concrete
- Presence and appearance of cracks
- Exposed rebar and signs of damage to coatings, sealings, expansion joints, etc.
- Deformation of the structure
- Presence of scour and erosion
- Upstream and downstream blockages
- Extent of marine growth colonization

The shortcomings of visual inspections when carried out by trained engineers in isolation—without any supplementary tools—have been well observed in the literature. Agin (1980) and Komorowski and Forsyth (2000) found that the quality of the assessment largely depends on the ability of the inspectors to observe and objectively record details of defects. Factors such as operator boredom, lapses in concentration, subjectivity, and fatigue contribute to the variability and reduced accuracy of visual inspections for terrestrial (topside) structures. These problems should be even more profound for underwater inspections as divers have to also compete with cold, uncomfortable, and hazardous conditions, as well as having to continuously monitor their air supply.

Gallwey and Drury (1986) found that incorporating image-processing techniques into the visual inspection regime offered far greater reproducibility over conventional visual inspection techniques. Utilizing image-processing techniques requires little additional effort as visual inspections almost always capture photographs to include in the inspection report to corroborate the inspector's comments; however, these photographs are rarely exploited to their complete potential in either a qualitative or a quantitative fashion. Furthermore, adopting effective image-based techniques can provide accurate quantitative information with minimal human supervision.

The main drawbacks associated with visual inspections and image-processing methods are the requirement of good visibility and the lack of penetration below the surface. Additionally, the presence of bio-fouling can inhibit visual inspections by masking the underlying structure, and in such cases, it must be cleaned beforehand. When subsurface defects need to be analyzed, inspectors can call upon a range of other non-invasive NDT tools.

2.3.4 Underwater non-destructive testing (NDT) tools

Choosing a suitable NDT tool for a particular inspection requires consideration of a number of factors and will depend on the on-site operating conditions and the type of damage present on structures. These factors are presented in Table 2.2.

In the highly corrosive marine environment, common forms of damage include chloride-induced corrosion and cracks that form on concrete

Table 2.2 Key factors when choosing an NDT tool

Factor	Reasoning
Cost	The cost of sourcing and operating NDT tools that can be deployed underwater is a major factor.
Sensitivity	Sensitivity is required to properly detect and characterize defects.
Speed	Speed is an important economic factor, and divers can only spend a limited time underwater.
Coverage	The ability to assess larger portions of the structure means that defects are less likely to be missed and the inspection can proceed more efficiently.
Reliability	Reliability has important cost implications since reporting defects can lead to early intervention before problems escalate.
Portability	NDT tools that are easy to maneuver by divers, or can be carried by ROVs, have significant practical benefits.
Versatility	The ability of an NDT tool to detect various damage forms is advantageous.
Training	NDT tools that require minimal training reduce costs and offset the need to hire highly trained and experienced individuals/divers.
Ease of operation	Tools that are straightforward to operate decrease the risk of obtaining erroneous measurements, which can undermine an inspection.
Safety	Certain tools are dangerous underwater, e.g., x-ray methods emit radiation that is particularly harmful underwater. Additionally, tools that require intense focus can cause divers to neglect their own safety.

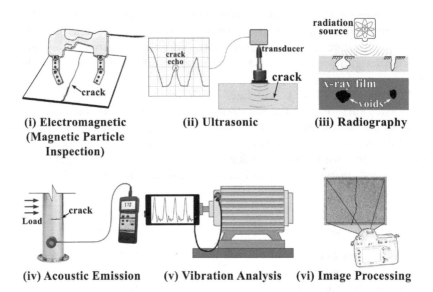

(i) Electromagnetic
(Magnetic Particle
Inspection)

(ii) Ultrasonic

(iii) Radiography

(iv) Acoustic Emission

(v) Vibration Analysis

(vi) Image Processing

Figure 2.5 (i–vi) Common NDT tools.

structures due to volume expansion of corroding reinforced steel. Image-processing is well-suited for detecting these damages; however, there are a range of other underwater NDT techniques that offer various capabilities.

Underwater NDT techniques are adapted from top-side NDT techniques. The modifications usually entail waterproofing and tuning the instruments to suit the underwater environment. Common underwater NDT techniques are (i) Electromagnetic, (ii) Ultrasonic (US), (iii) Radiography, (iv) Acoustic Emission (AE), and (v) Vibration Analysis based, which are depicted in Figure 2.5. The capabilities and the drawbacks of each of these techniques are explained in this section.

2.3.4.1 Electromagnetic methods

Electromagnetic methods are the second most popular method after visual inspections (Rizzo, 2013). Electromagnetic methods include magnetic particle inspection (MPI), eddy currents, electrical potential techniques, alternating current field measurement (ACFM), and magnetic flux leakage techniques. These techniques provide information about surface and near-surface defects, and about the effectiveness of cathodic protection systems for metallic structures. MPI methods are commonly used underwater to detect surface and near-surface flaws in ferromagnetic materials. They are capable of defining the true length of discontinuities. MPI methods can only be applied to bare metal surfaces and, since the process is time-consuming, they are generally only applied in small areas that have been identified as being susceptible to cracking (Lindgren et al., 2002). ACFM techniques are non-contact electromagnetic techniques that offer the capability to detect and size surface-breaking cracks in a range of different materials and through coatings of varying thickness. ACFM methods can also be used for the purpose of evaluating marine growth thickness. This task requires a diver to push a probe against the marine growth for a few seconds. The distance between the probe and the face of the underlying metal structure is then determined. While this approach is accurate and straightforward, the output is a series of spot measurements, which is not as intuitive to interpret and visualize as image-based approaches.

2.3.4.2 Ultrasonic methods

Ultrasonic methods can be applied to a wide range of materials and offer the capability to detect both external and internal defects, thickness measurements, and weld examinations (Rose et al., 1983). US methods are based on the propagation of ultrasonic waves generated by one or more probes through the structure. They are essentially local methods and therefore would require a lot of time if applied to large structures.

2.3.4.3 Radiographic methods

They can detect internal flaws in any material (Correa et al., 2009). Radiographic methods have the advantage of being accurate, the output is easy to interpret and visualize, and they have good resolution. However, the major drawback of radiographic methods is that there are inherent safety concerns when deployed in an underwater environment due to the emitted radiation.

2.3.4.4 Acoustic emission

Acoustic emission tools can be used to monitor the progression of damage, estimate the extent of corrosion in reinforced concrete structures, and detect cracks (Blitz & Simpson, 1996). Acoustic emission methods are attractive for monitoring large structures such as offshore platforms. The continuous wave and current loads that act on these structures provide an external stimulus that causes stress waves to propagate throughout the structure. Detection and analysis of the acoustic emission signals can reveal the location of any defects, as well as offer a qualitative assessment of their severity. However, the defect may remain undetected if there is not sufficient loading to produce an acoustic event.

2.3.4.5 Vibration analysis

Vibration analysis techniques rely on the natural vibration modes. In the marine environment, vibrations are typically induced by wind and waves. Like acoustic emission methods, vibration analysis techniques are well-suited for monitoring large structures; a major drawback is that they cannot determine the presence of damage until it becomes large enough to affect the natural vibration frequency of the structure. Moreover, environmental factors such as temperature must be accounted for, and the precise location of a broken member is not easy to determine.

2.4 CONVENTIONAL PHOTO COLLECTION PROCEDURES

Inspection companies and infrastructure management bodies sometimes have their own guidelines on how to collect photographs for inspectors in the field. The main purpose of these recommendations is generally to facilitate organizing and archiving the photographs and outlining what parts of structures should be photographed rather than specifying any technical aspects of imaging. Common recommendations include the following:

- Ensuring the date and time are properly set on the camera
- Including an item such as a pencil or ruler in the scene to help the viewer gain an appreciation of scale when photographing various structural elements (Figure 2.6a)
- Taking a photo of the structure identification to help recognize the photos later

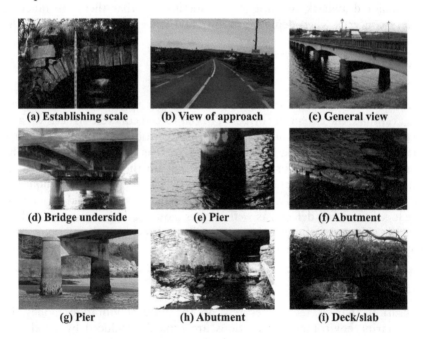

(a) Establishing scale   (b) View of approach   (c) General view

(d) Bridge underside   (e) Pier   (f) Abutment

(g) Pier   (h) Abutment   (i) Deck/slab

Figure 2.6 (a–i) Example inspection photographs.

Depending on the type of structure, there may be a checklist of components to photograph. In the case of a bridge, for example, inspectors may be requested to take a photo of the approaches, bridge surface, parapets, expansion joints, abutments, piers, bearings, wingwalls, the deck, and the underside of the superstructure. It is also customary practice to take a single general view photograph of the structure that shows as much of the structure as possible, or multiple photographs if the structure looks different from various sides. Some typical inspection photographs are shown in Figure 2.6.

A photo log can help inspectors keep track of what has been photographed and can be used to add descriptions of damage or other remarks on condition. Inspectors can link photographs to more detailed damage descriptions by referencing inspection notes. Keeping a log is useful when compiling the inspection report, especially for larger structures or when many components must be photographed. A sample photo log is shown in Figure 2.7.

In conventional inspections, imagery is typically acquired on an ad-hoc basis. This results in a large variability in terms of the viewing angles, camera-subject distances, perspective difference, exposure levels, etc. There is no agreed protocol for image collection and subsequent interpretation within the inspection framework (Phares et al., 2004). This book aims to partly fill that void by detailing a full and comprehensive set of best practice guidelines. By following a more systematic approach for acquiring images on-site, the input images become more consistent and more amenable for quantitative imaging.

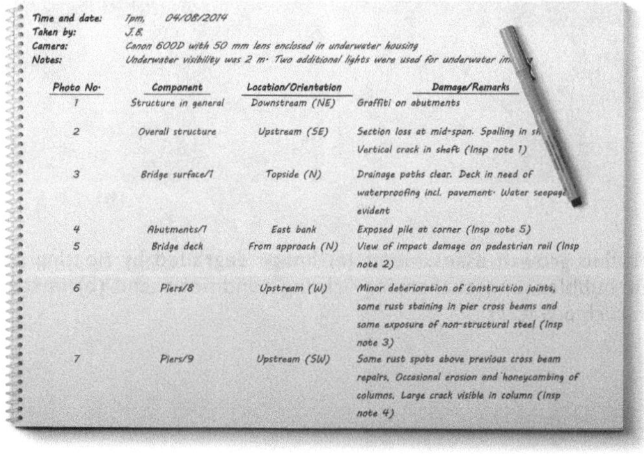

Figure 2.7 Example of a photo log from an inspection routine.

2.5 UNDERWATER PHOTOGRAPHY

While image-processing methods have the potential to be a convenient and useful underwater NDT tool, the poor underwater visibility conditions diminish the ability of cameras, and subsequent image processing techniques, to effectively assess damage. Underwater images are typically affected by issues such as light attenuation, scattering, color absorption, suspended particles, and air bubbles. These issues lead to blurriness, reduced contrast, and an overall loss of image quality. This problem is exacerbated in high turbidity conditions or when intense artificial light sources are employed. Such lighting creates highly non-uniform lighting, which frequently results in high specular reflections that mask details and creates bright spots that could mislead damage detection algorithms. Examples of some of these issues are shown in Figure 2.8. The image in Figure 2.8(a) was captured in the Mediterranean Sea as part of an inspection to measure the extent of marine growth colonization. The image is severely compromised due to the presence of air bubbles, floating particulate, and the harsh lighting conditions. The image in Figure 2.8(b) depicts a pier in the River Lee in Cork, Ireland, that is also affected by bio-fouling. In this case, the turbid waters reduce visibility and make it more difficult to observe detail.

Apart from devising protocols for better data collection, there is also an emphasis on devising techniques that can handle the underwater conditions with credibility. This book outlines methods for detecting cracks, surface damages, and obtaining 3D shape information in challenging underwater scenes.

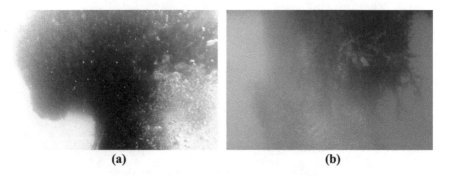

(a)   **(b)**

Figure 2.8 Marine growth assessment: (a) image degraded by floating particulate, air bubbles, and challenging lighting conditions, and (b) image captured in turbid waters.

2.6 SCOPE FOR INTEGRATING IMAGE-BASED
TECHNIQUES INTO INSPECTIONS

Images are captured routinely in most built infrastructure assessments. However, there exists significant scope for further utilization, as currently, the full potential of cameras is not being fully exploited. Cameras are a convenient and versatile tool capable of making visual data a part of quantitative assessment for a wide array of applications in Structural Health Monitoring (SHM). For inspectors to properly capitalize on the power of cameras and integrate them into the inspection framework, they must be aware of the camera's limitations, know how to acquire photographs effectively, and have access to efficient image processing techniques.

There are considerable advantages of using image-based methods as part of the inspection process. The extensive effort and expense associated with undertaking inspections warrant the development of sophisticated image-based techniques that can fully exploit the available scene information. This should be accompanied by a thorough exploration of their on-site performance levels so that inspectors are equipped with tried and tested NDT methods.

The value of image-processing lies in its simplicity and its versatility. Compared with other techniques, it is more intuitive for humans to interpret the visual data. This is especially useful in the case of offshore inspections as the imagery obtained from the cameras can be easily reviewed on-site. If it is apparent that the image quality is subpar (for instance, the imagery is too dark or there is excessive motion blur), the inspection can be repeated as necessary using different camera settings or additional light sources. For other methods, the inspectors may not know what they are looking at or what it relates to on the structure when they are back at the office. Additionally, it is often at this stage when problems with the data are realized for the first time. Moreover, cameras are readily available and require only minimal training, which is also in contrast with other NDT techniques.

The versatility of image-processing solutions is demonstrated by their ability to assess the most common damage forms affecting marine structures. They can even be used as a precursor for other techniques whereby areas that are likely to be damaged are initially identified based on visible surface anomalies (such as surface discoloration in the case of pitting corrosion). These areas are then elected for further scrutiny from localized NDT tools that perform more detailed subsurface examinations.

This book aims to improve the integrability of image-processing methods into the overall SHM paradigm. First, new techniques are developed to address the unique problems encountered in an underwater infrastructural setting, and existing image-processing techniques are customized to suit SHM applications. Second, a comprehensive image repository is presented, and an image acquisition protocol is established so that inspectors are equipped with the necessary knowledge to carry out camera based inspections.

2.7 USING IMAGE-PROCESSING DATA FOR SUBSEQUENT ANALYSIS

The quality of inspections can be enhanced by adopting efficient and dependable methods for acquiring, integrating, and interpreting structural performance data for maximum useful information at a minimum cost while offsetting or supplementing the qualitative, subjective, and unreliable human element. Incorporating image-processing techniques into the inspection regime serves this purpose as it counteracts the inherent limitations of visual inspections. For example, capturing the 3D shape of underwater structural components, instead of just a 2D projection as a standard camera does, naturally lends itself to numerous applications. This opens the possibility of performing a host of subsequent analysis tasks such as computational fluid dynamics (CFD) simulations using the reconstructed 3D geometries. Moreover, once engineers have access to scaled metric 3D reconstructions of underwater targets, physical properties can be easily extracted, such as the size and shape characteristics damage forms like cracks and corrosion. Detection of these damages may be easily automated using image-based damage detection techniques described in this book.

While it is important that human error from visual inspections is reduced, it is also important that the input information from an NDT technique is accurate and comprehensive. The measure of the onsite performance of an NDT tool remains a pertinent question in the majority of cases (Schoefs et al., 2012). The most suitable NDT for a given application will not only depend on the damage to be detected but will also depend on the operating environment. Applying NDT techniques in an underwater environment typically results in a decline in performance compared to top-side application in ideal conditions. This is similarly true for image-based techniques.

While there has been a relatively great deal of attention devoted to developing image-based techniques, very little follow-up work has been done to map their performances under varying operating conditions. This book describes a unified framework in Chapter 8 that offers the capability of assessing the performance of any image-processing technique in an underwater setting. This would enable inspectors to decide on the feasibility of adopting an image-processing based approach prior to inspection and to isolate conditions that are conducive to good performance.

2.8 CONCLUSION

This chapter highlights the importance of carrying out inspections, especially in relation to marine structures. The motivations for carrying out cost-effective inspections are discussed, and the financial implications of effective monitoring are detailed. An overview of the traditional inspection process for marine structures is provided and areas where there is scope for improvement are identified. The efficacy of inspections is only as good as the ability of inspectors, and the NDT tools at their disposal, to reveal defects and track the progression of damage with time. Hence, there is value in augmenting visual inspections with additional quantitative data that can help to this end.

There seem to be significant benefits of utilizing image-processing based methods as part of the inspection regime. It is seen that such methods are well-suited for underwater application, and ultimately, they have the potential to be a powerful tool for the cost-effective safety management of marine structures.

There are several opportunities for integrating image-processing based methods into the inspection framework effectively and in a more systematic fashion.

REFERENCES

Agin, Gerald J. "Computer vision systems for industrial inspection and assembly." *Computer 5* (1980): 11–20.

Blitz, Jack, and Geoff Simpson. *Ultrasonic methods of non-destructive testing.* Vol. 2. Springer Science & Business Media, 1995.

Boéro, Jérôme, Franck Schoefs, H. Yáñez-Godoy, and Bruno Capra. "Time-function reliability of harbour infrastructures from stochastic modelling of corrosion." *European Journal of Environmental and Civil Engineering* 16, no. 10 (2012): 1187–1201.

Busby, F. R. "Underwater inspection/testing/monitoring of offshore structures." *Ocean Engineering* 6, no. 4 (1979): 355–491.

Chandler, Kenneth A. *Marine and offshore corrosion: Marine engineering series.* Elsevier, 2014.

Correa, S. C. A., E. M. Souza, D. F. Oliveira, A. X. Silva, R. T. Lopes, C. Marinho, and C. S. Camerini. "Assessment of weld thickness loss in offshore pipelines using computed radiography and computational modeling." *Applied Radiation and Isotopes* 67, no. 10 (2009): 1824–1828.

Design Manual for Roads and Bridges, Highway Structures: Inspection and Maintenance Assessment, BD 21/01, vol. 3(4), 2001.

Det Norske Veritas. "Risk management in marine and subsea operations, Offshore standard DNV-RP-H101." DNV, Norway, http://exchange.dnv.com/OGPI/OffshorePubs/ViewArea/RP-H101.pdf (2003).

Gallwey, T. J., and Colin G. Drury. "Task complexity in visual inspection." *Human Factors* 28, no. 5 (1986): 595–606.

Gudmestad, Ove T. "Challenges in requalification and rehabilitation of offshore platforms—On the experience and developments of a Norwegian operator." *Journal of Offshore Mechanics and Arctic Engineering* 122, no. 1 (2000): 3–6.

Komorowski, Jerzy P., and David S. Forsyth. "The role of enhanced visual inspections in the new strategy for corrosion management." *Aircraft Engineering and Aerospace Technology* 72, no. 1 (2000): 5–13.

Lindgren, A., P. J. Shull, K. Joseph, and D. Hagemaier. "Magnetic particle." *Nondestructive Evaluation: Theory, Techniques, and Applications* (2002): 193–259.

Moan, Torgeir. "Safety of offshore structures." *Center for Offshore Research and Engineering, National University of Singapore, Singapore, Report* 4 (2005).

Petrequin, Marc. "The crisis in US wastewater infrastructure." (2011). Retrieved from http://www.academia.edu/1786909/The_Crisis_in_U.S._Wastewater_Infrastructure.

Phares, Brent M., Glenn A. Washer, Dennis D. Rolander, Benjamin A. Graybeal, and Mark Moore. "Routine highway bridge inspection condition documentation accuracy and reliability." *Journal of Bridge Engineering* 9, no. 4 (2004): 403–413.

Ramos, E. J. "Underwater inspection—The state of art." In *Non-destructive Testing 92*, C. Hallai, and P. Kulcsar, eds., Elsevier, Oxford, pp. 517–519, 1992.

Rizzo, P. "NDE/SHM of underwater structures: A review." In *Advances in Science and Technology*, vol. 83, pp. 208–216. Trans Tech Publications, 2013.

Rose, J. L., M. C. Fuller, J. B. Nestleroth, and Y. H. Jeong. "An ultrasonic global inspection technique for an offshore K-Joint." *Society of Petroleum Engineers Journal* 23, no. 2 (1983): 358–364.

Rouhan, A., and Franck Schoefs. "Probabilistic modeling of inspection results for offshore structures." *Structural Safety* 25, no. 4 (2003): 379–399.

Schoefs, F., Boéro, J., Clément, A., and B. Capra. "The αδ method for modelling expert judgment and combination of NDT tools in RBI context: Application to marine structures." *Structure and Infrastructure Engineering: Maintenance, Management, Life-Cycle Design and Performance (NSIE). Monitoring, Modelling and Assessment of Structural Deterioration in Marine Environments* 8(Special Issue), (2012): 531–543.

Stacey, A., M. Birkinshaw, and J. V. Sharp. "Life extension issues for ageing offshore installations." In *International Conference on Ocean, Offshore and Arctic Engineering. OMAE*, vol. 57411 (2008).

Wall, M., F. A. Wedgwood, and W. Martens. "Economic assessment of inspection—The inspection value method." *NDT. net* 3, no. 12 (1998).

Watson, J. "Underwater visual inspection and measurement using optical holography." *Optics and Lasers in Engineering* 16, no. 4–5, (1992): 375–390.

Chapter 3

Fundamentals of image acquisition and imaging protocol

3.1 INTRODUCTION

The maximum possible information that can be extracted from image analysis depends not only on the effectiveness of the image-processing algorithms but also on the quality of the input images. Quality and consistency are two important factors of image acquisition that need to be understood before developing any algorithms. This chapter discusses three areas that have a key influence on image quality and consistency:

- The camera hardware (principally the sensor and the lens), and the camera's general functionality and feature-set (e.g., having the ability to record in 4K resolution).
- The choice of camera settings (focus, exposure, recording format, etc.). This choice can typically be made manually by the photographer or left for the camera to automatically decide on the best settings for a given scene.
- The creation of good photographing conditions, such as providing ample illumination in the scene, and adopting effective image acquisition procedures such as steady handling of the camera.

3.2 CAMERA

Cameras come in many different forms such as compact digital cameras, DSLRs (Digital Single-Lens Reflex), mirrorless cameras, action cameras, and smartphones. Compact digital cameras, also known as point-and-shoot cameras, are affordable, easy to use, lightweight, convenient, and they do not require additional lenses. However, some drawbacks are that they produce noisier images due to their smaller sensors, they have limited aperture and zoom range, and they often do not allow users to fully control settings such as the shutter speed and aperture.

DSLR cameras are capable of delivering superior image quality and much better low-light performance, primarily due to their larger sensor sizes compared with compact digital cameras. DSLR camera bodies can be paired with a wide variety of interchangeable lenses to suit the requirements of different situations. DSLRs tend to be more expensive and heavier than other camera types, and they require know-how to operate effectively. Mirrorless cameras are essentially the same as DSLRs except that they have an electronic viewfinder as opposed to an optical one. This difference means that mirrorless cameras can be smaller than DSLRs.

Action cameras are small, rugged, and user-friendly. They are often waterproof or come with waterproof housings, which allows them to be readily used underwater. They have a wide field of view, so a lot of the scene is captured. Much like compact digital cameras, action cameras have small sensors that limit their low-light capabilities and lead to noisy images. Moreover, these cameras suffer from radial distortion, which introduces inaccuracies when analyzing and extracting measurements from imagery.

Smartphone cameras have progressed significantly over the years, and they now produce perfectly acceptable images. The advantage of smartphone cameras is that they are always at-hand and the captured images can be shared and transmitted easily. Nevertheless, the highest possible image quality will only be achievable by using a dedicated camera, such as a DSLR or a mirrorless camera.

Finding the right camera will inevitably be a compromise between price, convenience, and image quality. Image quality is quite a general term in this sense and covers several quality-related attributes such as sharpness, noisiness, brightness, resolution, contrast, and dynamic range. The main camera components that affect each of these attributes are the sensor and the lens. These are discussed in the following subsections.

3.2.1 Sensor

The sensor collects light energy and converts it into an electrical signal at every photo-site location. The camera's onboard computer chip then stores each electrical signal as a digital value that is proportional to the amount of light energy arriving at the sensor. The final image file is simply a collection of all the digital values, with some in-camera processing applied. The sensor has a considerable impact on image-quality related factors such as the image resolution, low-light performance, depth of field, and dynamic range. The main characteristics of the sensor that determine ultimate image quality are the physical size of the sensor, the pixel count, the dynamic range, and the underlying sensor technology.

3.2.1.1 Sensor size

A camera with a physically bigger sensor tends to capture a higher quality image by gathering more light but at the cost of larger optics leading to bulkier camera units. Small sensors, which are often found in action

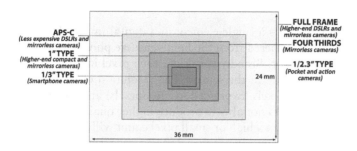

Figure 3.1 Common sensor sizes and the types of cameras where each sensor is often found.

cameras and smartphones, can be paired with smaller optics, which means that the overall size of the imaging system remains compact. Some common sensor sizes are shown in Figure 3.1. It is worth mentioning that all sensors within the same category do not have the same dimensions; the measurements provided are an example of one such sensor within that format.

The size of full-frame sensors is based on the traditional 35 mm film camera, approximately 36 × 24 mm, or 864 mm². This format is mostly found in high-end DSLR and mirrorless cameras. This sensor size provides the point of reference for describing lens length in mm using a term known as the "crop factor." The crop factor for a full frame sensor is 1.

There will be a different field of view when lenses of the same length are used in conjunction with sensors of varying size. The crop factor is multiplied by the lens length to reveal the "35 mm equivalent" field of view, that is, the field of view that would be produced if using a given lens with a full frame sensor.

There are several variations of APS-C sized sensors, but the most common size is 23.67 × 15.7 mm. These have a crop factor of around 1.5, meaning that a 35 mm lens mounted on an APS-C body becomes the equivalent of an FF "normal" lens at 35 × 1.5 = 52.5 mm ("normal" lenses are discussed in more detail in Section 3.2.2). Many mid-range DSLR and mirrorless cameras use this format. Micro four-thirds (MFT), 17.3 × 13 mm, or 225 mm², is an increasingly popular format that is used in some mirrorless camera systems. It is a compromise that permits cameras and lenses to be made smaller while still maintaining relatively good image quality. The same can be said for the slightly smaller 1-inch sensors (the diagonal length is approximately 1 inch).

The sensors found in many pocket cameras, action cameras, and smartphones vary in size, but popular ones include 1/2.3-inch and 1/3-inch sensors. The primary advantage of small sensors is the implications on size, weight, and cost of the overall camera unit. For many users, the practical benefits of a small camera system will outweigh the improvements in image quality that are obtainable with larger sensors.

3.2.1.2 Pixel count

Each pixel represents a single element of a digital image. The information for each pixel is gathered from the light-sensitive photosites on the surface of the camera sensor. Broadly speaking, each pixel in an image has a corresponding photosite on the sensor. The resolution of the image depends on the number of pixels (expressed in megapixels) and the quality of each pixel. More pixels do not always produce higher quality images.

The choice between a high or small pixel count will largely depend on the intended application. For instance, a 24MP megapixel image captured with an APS-C camera may outperform a 12MP full-frame camera if the application requires fine details to be resolved, such as cracks that might be only a few pixels wide. Conversely, a 12MP full-frame camera should be much better at identifying large corrosion stains in low-light underwater conditions. Moreover, for many computationally intensive image processing tasks, the input images will be downsized to facilitate computational efficiency. In such cases, there will be no value in capturing high megapixel images.

One noteworthy benefit of cameras with more megapixels is that the extra pixels will be helpful when cropping. Sometimes, there may only be a small region of interest within the image, or there may a lot of distortion or color aberrations near the fringes of the images (typical of wide-angle lenses) so the only usable part of the image is the central region. With more megapixels, the region of interest can be cropped and still retain a high degree of detail.

3.2.1.3 Dynamic range

The size of the sensor and the pixel count have a combined effect on the dynamic range. Dynamic range describes the range between the darkest and brightest points in a scene that a given digital camera can capture. The range of light values that human eyes can distinguish, from dark to light, is far greater than any sensor pixel can record. Any shadow value darker than what the sensor can detect is simply rendered as an effective black, and any highlight value brighter than what the sensor can detect is rendered as white. Generally, it can be assumed that cameras with larger photosites, or a greater pixel size, will have the ability to record a greater dynamic range. Larger sensors with fewer pixels can be an indicator of a larger photosite. Larger photosites allow for the collection of more light and, subsequently, more detail and a higher contrast ratio to be recorded.

3.2.1.4 Sensor technology

The size and pixel count are not the only factors that govern the image quality. The efficiency of the in-camera image processor and the type of sensor technology also play a role. Historically, there are two main types of sensors: CCD (charge-coupled device) and CMOS (complementary metal-oxide semiconductor). CCD sensors were widely used in video and stills cameras as they offered superior image quality to CMOS sensors, and they could achieve better dynamic range and noise control. However, the image quality of CMOS sensors has improved markedly in recent years, and their ability to work more efficiently and consume less power has seen them largely replace CMOS sensors on new cameras, in particular for cameras that offer high-speed capture functionality. There are several online resources that rank various cameras (and lenses), such as https://www.dxomark.com, which can help guide purchasing decisions.

3.2.2 Lens

The function of the lens is to depict the complete field of view on the sensor with the highest resolution, highest contrast, and smallest optical errors as possible. Lenses are chiefly described by their focal length and their maximum aperture, which will be discussed in this section.

3.2.2.1 Focal length

The focal length is the distance from the focal point of the lens to the plane of the sensor. Most lenses can be described as normal, wide-angle, or telephoto based on their focal length. The field of view of a typical normal, wide-angle, and telephoto lens is shown in Figure 3.2. Wide-angle lenses show a large portion of the scene, which can help viewers to better interpret the scene by providing more context. Telephoto lenses, on the other hand, provide detailed close-up views. It is important for inspectors to think about

Camara view with various lenses:

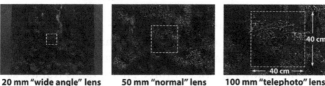

20 mm "wide angle" lens   50 mm "normal" lens   100 mm "telephoto" lens

Figure 3.2 The field of view of a 20 mm, 50 mm, and 100 mm lens on a full-frame sensor and how a 40 cm × 40 cm area, which is 2 m from the camera, is depicted by each lens.

how they wish to frame their photographs (i.e., how much of the scene they wish to capture) so that they can strike the right balance between capturing sufficient detail while also capturing enough of the scene.

A focal length of about 50 mm on a full-frame sensor is considered to be normal since it approximates the human field of view (about 40°). For this reason, normal lenses are often good for general inspection photography as they appear natural to human observers. Focal lengths shorter than normal are considered "wide angle," while longer focal lengths are considered "telephoto," all with reference to the full frame sensor. Wide-angle lenses can be moderate (24 to 36 mm) or as extreme as 17 mm or less. For underwater photography, it is worth noting that many photographs must be taken from much closer distances given the limited underwater visibility. Wide angle lenses are particularly applicable here as they allow a reasonably sized area to be covered when shooting at close range. Moreover, because of the refractive property of the water, underwater the apparent distances are about three-fourths of actual distances and objects appear larger than what they actually are. Ultra-wide-angle lenses are sometimes referred to as fisheye lenses, and these usually cover up to a 180° field of view. Fish-eye lenses produce strong visual distortion—instead of forming a normal rectilinear image, they form the kind of circular image that the name implies. Such images are not ideal for image-processing tasks as it is generally difficult to extract quantitative information given the high level of distortion, although this distortion can be partially removed if the lens profile is known (common lens models are available online) and remapped into a rectilinear image; however, this process is another source of errors.

Longer focal lengths correspond to higher magnification and can also be moderate (90 to 135 mm) or super telephoto (over 300 mm). Telephoto lenses are useful for situations where it is not possible to get close-up access to the subject and yet there is a requirement to capture the subject in a lot of detail. An example of this would be photographing a section at the mid-span of a bridge with a camera that must be positioned relatively far away on the river bank.

The effect of sensor size is presented in Figure 3.3, where two photographs are taken with lenses of the same focal length (20 mm), but with a full-frame (FF) and an APS-C format sensor. Remembering that the crop factor of an APS-C format camera is 1.5, the view using a 200 mm lens on an APS-C sensor is the equivalent view as using a 30 mm (20 mm × 1.5) on a full-frame sensor.

Different focal lengths also change the perspective of an image: wide angle makes close objects seem much larger than farther ones while telephoto makes objects appear "flatter" or closer together, as depicted in Figure 3.4, which shows a series of three photographs framed as similarly as possible with three different lenses. Note that the distance between wind turbines appears greatest with the wide-angle lens. With each succeeding photograph, the wind turbines appear closer together. The normal lens is the best representation of how a human observer would perceive the scene.

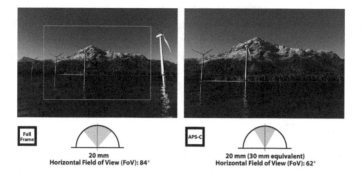

Figure 3.3 Same focal lengths on a full frame and on an aps-c sensor.

20 mm "wide angle" lens	50 mm "normal" lens	100 mm "telephoto" lens
(extension distortion)		(compression distortion)

Figure 3.4 Perspective changes between photographs taken with a 20 mm, 50 mm, and 100 mm lens, at different distances from the subject.

Given that different lenses produce different perspective effects, it is worthwhile to take note of the focal length of the lens when capturing imagery (if it is not automatically recorded in the EXIF metadata) so that viewers can better interpret the imagery.

3.2.2.2 Aperture

Aperture refers to the opening of a lens's diaphragm through which light passes. It is calibrated in f-stops and they are sequenced so that each value is approximately a multiple of $\sqrt{2}$, that is, 1, 1.4, 2, 2.8, 4, 5.6, and so on, although intermediate values such as 1.8 are also often encountered. A smaller f-stop indicates a larger opening of the diaphragm, which allows more light to reach the sensor, while a larger f-stop number indicates a smaller opening, which lets in less light.

When shooting, the camera can always be set to a higher f-stop to let in less light, but the smallest possible f-stop is governed by the maximum aperture of the lens. The maximum aperture of many lenses on the market is typically in the range from 1.6 to 5.6. In general, wider aperture lenses perform better in lower light conditions, but this comes with the side effect of having a shallower depth of field. While a shallow depth of field is often highly valued by non-technical photographers as it is seen as an aestheti-cally pleasing quality that helps isolate the subject from the background, it is not beneficial for the purpose of inspection photographing. For inspec-tion photography, it is desirable to get as much detail as possible from fore-ground to background, so choosing a smaller aperture (a higher f-stop, such as f/8 or f/11) provides a larger depth of field.

3.2.2.3 Types of lenses

Most action cameras and compact cameras have a built-in lens that allows for less expensive manufacturing, lighter and smaller cameras, and also reduces the risk of dust getting into the camera body. However, almost all DSLRs and mirrorless cameras employ an interchangeable-lens framework that provides much more versatility.

There are two main types of lenses: prime and zoom. Prime lenses only have a single, fixed focal length (e.g., a 50 mm lens). Zoom lenses, on the other hand, cover a range of focal lengths (e.g., an 18–55 mm lens). The key advantage of zoom lenses is their versatility, while the drawbacks are that zoom lenses have more lens elements, which mean less light reaches the sensor. Therefore, prime lenses are a better choice for low-light conditions. Moreover, prime lenses are often heavier, more expensive, and produce relatively inferior image clarity.

3.2.2.4 Other lens features

Two lenses with similar focal lengths and apertures can still differ greatly based on their features and build quality. The build quality and lens design will affect several lens properties such as sharpness, chromatic aberration, vignetting, and geometric distortion. In terms of features, two that are worth mentioning in the context of underwater photography (and photog-raphy in general) are image stabilization and autofocusing.

While top-side inspections can use a tripod as a steady base, underwa-ter photography must usually be collected with a handheld camera that is susceptible to vibrations, thus leading to blurry images, especially when shooting at slow shutter speeds as is often required in low light conditions. To partially address this problem, some modern lenses have incorporated a stabilization mechanism, although these are still far from perfect and, needless to say, it is best to have a steady hand when capturing imagery rather than relying on the lens's image stabilization. A stabilized lens is not needed if the camera system already offers in-camera image stabilization. In such a scenario, the stabilization mechanism on either the lens or the body should be disabled to prevent conflicts.

Manual focusing is a particularly arduous task to perform when under-water. Even for camera housings that allow the lens barrel to be turned, it is hard for the diver/photographer to accurately judge when the subject is in proper focus as the monitors on the back of cameras are not very clear when viewed underwater. Consequently, autofocusing becomes even more instrumental underwater as no image will be successful if it is not accurately focused. This does not mean that the entire image has to be sharp—only the most important parts. Autofocusing is a shared responsibility between the lens and the camera—most modern lenses are focused under control of the camera.

3.2.2.5 Filters and lens ports

Red filters are sometimes used for underwater photography as a way to reintroduce the red light that is lost due to color absorption (red colors are absorbed to a greater extent than blues and greens). Without a filter, the shift in available light will often confuse the camera's white balance metering. Red filter corrects for this and helps deliver more vibrancy and contrast to underwater photographs. However, red filters reduce the amount of light that reaches the camera sensor, so to compensate, there will be a need to shoot at higher ISO values, wider apertures, or slower shutter speeds. Overall, the improvements are small and red filters only help so much, especially considering that color-correction techniques can be applied to the imagery and achieve much of the same effect.

There are two main types of lens ports: flat and dome. Flat ports are not able to correct for the distortion produced by the differences between the indexes of light refraction in air and water. This leads to a number of issues in the imagery. First, the imagery appears magnified—the effective focal length of the lens becomes approximately 25% greater than the actual focal length. Second, because flat ports do not distort light rays equally, they have a progressive radial distortion that becomes more apparent as wider lenses are used. Light rays that pass through the center of the port are not significantly affected because their direction of travel is perpendicular to the water-air interface of the port. Finally, flat ports introduce some chromatic aberration, which is once again more noticeable with wider lenses. Hemispherical dome ports are designed to restore the focal range lost due to optical refraction, effectively adding about 33% wider field of view underwater. When a dome port is used, the rays of light pass through with minimal refraction, which allows the "in-air" lens to retain its angle of view. This reduces the problems of refraction, radial distortion, and axial and chromatic aberrations. High-quality dome lens ports can be quite expensive; however, they offer noticeable image quality benefits if working with wide angle lenses.

3.3 CAMERA SETTINGS

The camera settings have a major impact on image quality and, consequently, on the ability of image-processing algorithms to effectively identify instances of damage. The choice of cameras settings is normally a trade-off between obtaining an acceptably bright image while minimizing adverse quality factors such as blurriness and noise in the imagery. This section discusses the effect of each camera setting and offers some sensible conventions that can help inspectors to capture good quality images on a more consistent basis. Practical recommendations regarding which image file formats to use and how inspection imagery should be archived are also provided. Special attention is given to multi-camera systems as these systems require some additional considerations. Furthermore, in certain situations (i.e., static scenes), inspectors can use high dynamic range (HDR) imaging, which offers a way to capture more tonal detail from a scene. This section covers the settings that can be used to create HDR imagery.

3.3.1 Image archiving

The creation of an image library requires a set of predefined standards to ensure that all contributions are consistent. This forms the basis for an organized and manageable library. The proposed image archiving protocol discusses the file format, metadata and information to be recorded, and file naming convention and cataloguing.

There are a wide variety of image formats available. Typically, there is a trade-off between the image file size and the amount of information retained in it. As storage space is becoming increasingly inexpensive and data transfer rates between memory cards and computers are becoming faster, capturing images in both JPEG and RAW format (if this option is available) provides the advantages of both formats. JPEG images, which have a ".jpg" extension, are a universal standard for images and can be displayed by virtually any device. JPEG uses a method of lossy compression, which means that the image quality degrades after the image has been saved. The degree of compression can often be controlled by the photographer. Camera menus might use adjectives such as "medium" and "fine" or present the user with a numerical scale. The highest quality possible should be selected; it can always be reduced later. At the highest quality setting, image degradation is barely perceivable. The more manageable and smaller-size JPEG images are still well-suited for conventional image processing and analysis tasks, especially for tasks that require many images to be loaded at once, which, if larger files were used, could strain computational resources.

The RAW image format should be retained for archival purposes. RAW images contain minimally processed data from the image sensor. They are capable of storing a maximum level of information from a scene (i.e., wider dynamic range and color gamut). Post-processing of RAW files allows for all adjustment of exposure, color, and other qualities of the image. The RAW format differs depending on the camera model and camera manufacturer, although each format contains essentially the same data and metadata.

An additional step, if desired, would be to convert the original RAW formats to an open standard and well-supported format, namely the digital negative (DNG) format, which is a popular and freely available format developed by Adobe Systems™. Since there is a wide range of proprietary RAW formats, it is hard for applications and programs to guarantee future compatibility with them all, especially for some of the lesser known and lesser used RAW formats. Thus, the additional step of converting to the DNG format from the original RAW format would be of particular value to users that have, or expect to have, imagery acquired from a number of devices, and would like to unify the RAW formats into a common format that retains all of the original information.

In the case of video, still frames may be extracted at relevant intervals and saved as JPEG files as analyzing every frame is often unnecessary. The intervals are primarily determined by the speed of the camera relative to the subject and the task at hand. If a recording device is moving quite quickly, then more frames should be extracted. Generally, extracting three frames per second should be sufficient for most cases. The original video should be stored in its native format.

Photographs and video have accompanying metadata that contains useful information about the content and context of the file. Metadata is automatically embedded into each digital file. It provides information such as the time and date of capture, camera model, exposure information, etc. It is vital that the time and date of all contributing cameras are precisely set as this provides a convenient way for identifying the synchronized stereo image pairs. Some metadata must be manually added such as the baseline distance for stereo imaging or the camera calibration data. The specific nature of this metadata will vary according to the task. In the case of stereo imaging, the imagery should be marked as coming from the left or right camera.

3.3.2 Focusing

Focusing is less of an issue for wide angle lenses (such as those found on small action cameras) as the large depth of field means that most (if not all) of the scene is in focus. For normal and telephoto lenses, however, focusing is a key issue. Thankfully, modern autofocus cameras and lenses make underwater photography much easier. Many cameras offer a number of focus points within the visible frame that can be used to set the focus. A good strategy for underwater photography is to set the autofocus to use only the center focus point and ensure that the camera is always centered on the subject. Autofocusing can fail at times, especially if there is not a lot of contrast in the scene—something that occurs regularly in underwater scenes. In such cases, the best option would be to (i) manually focus at a point a certain distance away (e.g., 2 meters); (ii) select "aperture priority mode" on the camera, which keeps the aperture constant; and (iii) choose a small aperture (a higher f-stop, such as f/8 or f/11), which provides a larger depth of field. The diver can then capture images of subjects that are approximately the same distance away underwater with a high degree of confidence that the subject will be in focus.

3.3.3 Aperture, ISO, and shutter speed

There are three elements that control exposure: aperture, shutter speed, and ISO, as illustrated in Figure 3.5. Each element affects the exposure value (EV) in a different way. As mentioned previously, aperture is the size of the lens opening, and it regulates the amount of light that reaches the camera sensor. A large opening allows more light to reach the sensor,

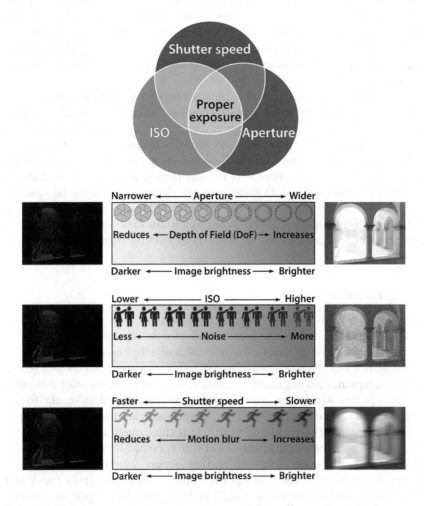

Figure 3.5 The elements that control exposure and their effect on image quality.

while conversely, a smaller opening restricts the amount of light reaching the sensor. Shutter speed controls how long the shutter is open, and thus controls the duration of light that is allowed to hit the camera sensor. The longer the shutter speed, the more light that falls on the sensor. ISO measures the sensitivity of the camera's sensor. The lower the ISO, the less sensitive the sensor is to light. The higher the ISO, the more sensitive it is.

Obtaining a properly exposed photograph means striking a balance between the aperture, ISO, and shutter speed. A wide aperture allows more light to reach the sensor at the expense of a narrow depth of field, meaning only objects within a confined range will be in focus. A small aperture will have a greater depth of field; however, the resulting image will be dark/underexposed unless the shutter stays open for an extended period of time to let enough light reach the sensor. Keeping the shutter open for too long (i.e., having a slow shutter speed) presents other problems, mainly a high degree of blur. Higher ISO settings tend to be used in darker situations to amplify the available light, but this comes at a cost of increased noise in the images. Many modern cameras have simplified the process of choosing the optimum aperture, ISO, and shutter speed settings for a given scene through various automatic exposure modes and the use of through-the-lens (TTL) metering. The camera will automatically attempt to counteract the dimly lit underwater conditions by combining a wide aperture with a slow shutter speed and a higher ISO. It is important that the settings remain within certain limits to minimize the impact of these problems. Recommendations of the limits are summarized in Table 3.1.

The shutter priority mode is recommended as diver-held cameras employed in underwater inspections are prone to shaking, which introduces motion blur. This is a semi-automatic shooting mode that allows the user to specify the shutter speed. The camera then automatically decides the best aperture and ISO sensitivity for the specified shutter speed to get the correct exposure. The shutter speed should not be any slower than 1/20 seconds. The imagery obtained from the cameras should be reviewed at regular intervals. If it is apparent there is too much motion blur present in the images, the shutter speed should be adjusted to a faster setting.

Additional artificial lighting will be required if the camera settings exceed any of these ranges. Irrespective of these requirements, artificial lighting will be necessary at greater depths where ambient lighting is not sufficient.

Table 3.1 Acceptable ranges for camera settings

Camera setting	Limit
Aperture	f/8–f/16
Shutter speed	As a rule of thumb, the minimum shutter speed is the inverse of the lens's focal length, i.e., minimum Shutter Speed (secs) = 1/Focal Length (mm).
ISO	An ISO value of 800 is a good compromise for photographing in dimly lit conditions while still controlling noise/graininess.

3.3.4 HDR

HDR imagery is a set of techniques that are used to allow a greater dynamic range of luminance values between the brightest and darkest regions of an image than standard digital images, and it has become an inbuilt feature in many cameras recently. SDR images can typically only accommodate a limited range bracket of the full tonal spectrum in a real-world scene. Therefore, a dynamic range bracket would have to be chosen in the knowledge that all luminance values outside the range would not be represented correctly. The broad principle behind HDR imagery is that multiple SDR images of the same scene, each taken at a different exposure, and thus capturing a different range bracket of the tonal spectrum, may be merged to form one HDR image that has a wider dynamic range (Reinhard et al., 2008). Combining SDR images can be done using various merging algorithms (Debevec & Malik, 1997; Naccari et al., 2005). In order for the SDR images to be aligned correctly, the camera must be mounted on a tripod or otherwise securely held between exposures.

The benefits of adopting HDR imagery as an imaging protocol may be observed in Figure 3.6, which shows three SDR images (an underexposed, a normally exposed, and an overexposed image) and the corresponding HDR image. These images depict a 30-year-old corroded steel pile in the tidal area in a wharf situated off the French Atlantic Ocean. The pixel dimensions of the images are 2816 pixels by 2112 pixels and the dimensions of the corroding metallic surface are approximately 0.3 m by 0.3 m. It may

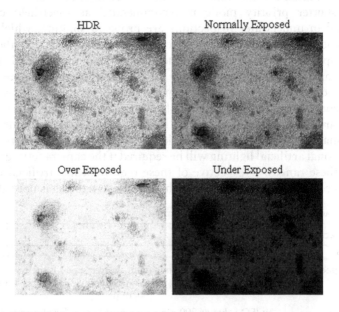

Figure 3.6 **A high dynamic range (hdr) image of corroding steel formed by merging the normally, over-, and underexposed images.**

be observed that HDR imagery is particularly useful here since the shiny metallic surface gives rise to naturally high dynamic ranges. Generally, scenes that have a wide dynamic range are likely to especially benefit from the adoption of HDR as a protocol.

3.4 GUIDELINES FOR OBTAINING GOOD QUALITY IMAGERY FOR QUANTITATIVE ANALYSIS

This section outlines a set of best practice guidelines for capturing photographs and video footage on-site. Adopting a well-defined and systematic approach for acquiring imagery helps to ensure that the imagery is consistent and, ultimately, more amenable to quantitative image analysis. Imagery that is well-suited for quantitative image analysis is characterized by a number of features. Naturally, image quality is important, but it is also important that the best viewing angles, camera-subject distances, etc. are selected when photographing the structural components under inspection.

3.4.1 Collection protocol

The diving protocol addresses the logistical considerations (testing equipment prior to usage, route planning, etc.), as well as handling lighting and turbidity conditions, which will affect the optimum distance between camera(s) and subject. It is vital that any unnecessary time spent underwater by the diver is kept to a minimum. With this in mind, the diver should be presented with a clear and concise brief outlining the task at hand. A rough idea of the turbidity conditions should be known beforehand. Both lighting and turbidity are crucial factors that affect the underwater visibility and consequently the image quality (Mahiddine et al., 2012). Artificial lighting in the form of underwater strobe lights or flashlights is required in dim lighting conditions.

3.4.1.1 Photographic lighting

The lighting for underwater photography is especially important because ambient light is often insufficient. Even in relatively clear water where natural light penetrates to the depth at which photo-documentation is to take place, artificial light sources should be used to obtain true color reproduction. Underwater lights come in a variety of forms. There are strobe lights, which are synced to the camera and send out a burst of light the moment that the camera takes a photograph, and there are flashlights, which remain on continually and can be used for videoing as well as capturing still photographs. One of the most important properties of the lights is their beam pattern. Lights can be categorized as a spotlight or a floodlight. Spotlight beam patterns are known for casting a narrow, long-reaching beam of light. Floodlights, on the other hand, cast a wide, all-encompassing wash of light that does not reach for a very long distance, as shown in Figure 3.7.

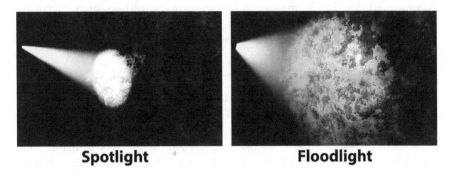

Spotlight   **Floodlight**

Figure 3.7 **Beam patterns: spotlight vs. floodlight.**

Floodlights provide more uniform light in the scene and are much less likely to cause "bright-spots," which mask surface details and may mislead image processing algorithms. Floodlights are particularly useful when using wide angle lenses as a lot more of the scene is captured and a wide light pattern is required to illuminate all of the photographed scene.

One issue to be mindful of when using artificial lights is backscatter. Suspended material reduces the light that actually reaches the subject, and it can reflect light back to the camera lens. To minimize this problem, lights should be placed as far away from the camera as possible so that light does not reflect directly into the lens. It is usually best to use two light sources of lower intensity located on each side of the camera.

3.4.1.2 Turbidity

Water is seldom optimally clear, and the dissolved and suspended matter can reduce visibility by both absorption and scattering of light. While turbidity may not be easily reduced, there are some precautions that can be taken to offset the deleterious effect in relatively high turbid waters. First, caution should be taken in shallow waters to avoid disturbance with the sea/river bed, which may unsettle fine sediments through either direct contact or from turbulence created from the ship. Anchoring the vessel downstream away from the inspection site can also prevent additional sediments that would reduce the water clarity. Second, the distance between the cameras and the subject under consideration should be reduced. In seas and oceans, the water is generally clear so measures to counteract poor visibility need not be considered.

Choosing the distance between the subject and the diver is a trade-off between a number of factors. The ideal distance should be kept in the range 50–150 cm; however, this will vary depending on the following circumstances:

- Visibility
- The size of the subject in the scene
- Acceptable error tolerance

When the underwater visibility is poor, the diver can photograph in close proximity to the subject (up to 30 cm). In extreme cases, a clearwater box may be employed. A clearwater box is a box constructed of clear acrylic plastic that can be filled with clean water through which the camera can be aimed. When the box is pressed against the subject, the turbid water is displaced, allowing the camera to focus through the space of clearwater. Various sizes and shapes of boxes are available, depending on the objects to be photographed. The use of the clearwater box normally requires two divers: one to operate the camera and one to help control the box.

3.4.1.3 Underwater stereo image acquisition

Stereo imaging requires a little closer examination than standard image acquisition. To begin with, stereo images should not be captured any closer than about 30 cm from the subject as this leads to large perspective differences between the stereo image pairs, which can hamper the matching process and result in myriad occluded regions (Matthies & Shafer, 1987). In clear underwater conditions, the diver can photograph from a distance up to 2.5 m before the stereoscopy breaks down and the error tolerance becomes unacceptably high (Olofsson, 2010).

If the object under inspection is quite large (e.g., a wide diameter pile) and it is wished to include the whole structure within an image, then photographing from farther back will provide better context, bearing in mind that the 2.5 m limit should not be exceeded. If, however, a greater accuracy of the structure's macro geometry is required, then the diver should not go farther than 1.2 m from the subject.

Finally, the baseline distance for stereo imagery will influence the choice of subject-camera distance. While the baseline shift will normally be fixed at a certain width according to the constraints of the available equipment, it is important to note the effect it has on the accuracy. Theoretically, a wider separation between the cameras results in a lower percentage error in the depth estimation. However, the advantages of having a wide separation are mitigated by the creation of large perspective differences as previously eluded to. The baseline shift should be in the range 10 cm to 30 cm. Additionally, the cameras should be aimed inward at an angle θ (known as the vergence angle) such that their center lines intersect approximately at the face of the subject as shown in Figure 3.8. This is to ensure that the cameras capture as many of the same points in both images as possible. The vergence angle has previously been used to model the error in depth by Sahabi and Basu (1996). For real-world scenes, they found that vergence angles in the range 5°–10° provided the lowest errors.

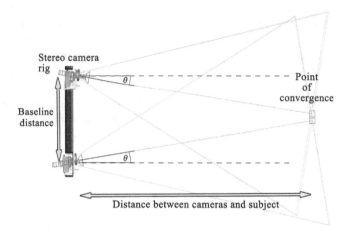

Figure 3.8 **Stereo rig set-up.**

3.4.1.4 Logistical considerations

All the equipment should be checked above water prior to each inspection. The underwater housings should be checked for any signs of leakage by first submerging them without their contents. If necessary, the O-rings should be relubed/replaced as per manual instructions. Care should be given to ensure that the time and date of the cameras are precisely set, that there is enough storage capacity in the SD cards, and the battery is sufficiently charged. The cameras, ideally, should be focused at a point on the subject as illustrated in Figure 3.8. Appropriate settings should be configured for each camera/video recorder, as previously discussed, ensuring that the shooting modes in both are identical. It is advised to initiate filming immediately before the diver submerges as it is easier to control the simultaneous triggering of both cameras when above water. The captured imagery should be reviewed at regular intervals during the inspection and any adjustments should be made accordingly. Dives that produce substandard imagery should be repeated. Necessary props should be prepared such as lighting equipment, an object of known dimensions to attach onto the structure, or a checkerboard for calibration if applicable. Moreover, this would be a good opportunity to record the baseline distance and the dimensions of the object of known size such as a checkerboard/wand. Finally, the diver should familiarize himself/herself with the blueprint of the structure and identify any components that are of particular interest. A suitable route should then be planned based on this groundwork. In cases where the diver cannot photograph a particular component from all sides due to restricted access, blockage by other obstacles, etc., he/she should endeavor to photograph as much of it as possible.

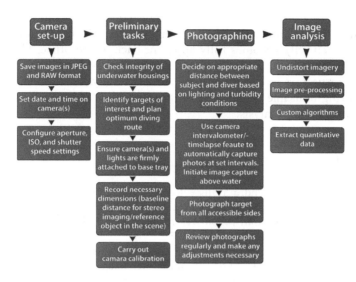

Figure 3.9 **Key steps involved in the stereo imaging pipeline.**

3.4.1.5 Combined underwater protocol

Greater care and attention is required for stereo imaging as two cameras must act in unison. A flowchart showing the summarized methodology for the combined underwater protocol for stereo imaging is shown in Figure 3.9.

3.5 CONCLUSION

The camera equipment and the image collection practices have a significant bearing on the quality and consistency of the obtained imagery. Capturing subpar inspection imagery means that the effectiveness of image-processing algorithms will be limited from the outset. Although image enhancement techniques can help to some extent to improve image clarity, inspectors should be mindful of the fact that no amount of digital post-processing can overcome flaws in the original out-of-the-camera file. There is thus an emphasis on using the right equipment and doing everything possible during the image acquisition stage to ensure that high-quality images are captured consistently and reliably.

While good camera equipment is essential for delivery of good image quality, the full extent to which a camera's possibilities are realized depends in part on the proficiency of the user; adopting effective image acquisition practices and choosing the right camera settings are just as important as having suitable camera equipment. This chapter presents a comprehensive protocol that sets out best practice guidelines for collecting underwater image and describes ways to minimize the likelihood of certain problems from occurring, such as out-of-focus or overly noisy imagery.

Adhering to a protocol provides reassurance to inspectors as it can often be hard to fully appraise the image quality on-site. The true quality of the imagery often only becomes fully apparent back at the office. Having an understanding of the camera settings, and how they impact the imagery, can help inspectors to choose settings more carefully. This lessens the risk of capturing unusable or substandard imagery that is not well-suited for quantitative image analysis.

REFERENCES

Debevec, Paul E., and Jitendra Malik. "Recovering high dynamic range radiance maps from photographs." In *Proceedings of the 24th annual conference on computer graphics and interactive techniques*, pp. 369–378. ACM Press/ Addison-Wesley Publishing Co., 1997.

Mahiddine, Amine, Julien Seinturier, Daniela Peloso Jean-Marc Boi, Pierre Drap, Djamel Merad, and Luc Long. "Underwater image preprocessing for auto-mated photogrammetry in high turbidity water: An application on the Arles-Rhone XIII roman wreck in the Rhodano river, France." In *Virtual Systems and Multimedia (VSMM), 2012 18th International Conference on*, pp. 189–194. IEEE, 2012.

Matthies, Larry, and Steven A. Shafer. "Error modeling in stereo navigation." *IEEE Journal on Robotics and Automation* 3, no. 3 (1987): 239–248.

Naccari, Filippo, Sebastiano Battiato, Arcangelo Bruna, Alessandro Capra, and Alfio Castorina. "Natural scenes classification for color enhancement." *IEEE Transactions on Consumer Electronics* 51, no. 1 (2005): 234–239.

Olofsson, Anders. "Modern stereo correspondence algorithms: Investigation and evaluation." PhD thesis, Linkoping University, Sweden, 5–86 (2010). Retrieved from http://liu.diva-portal.org/smash/get/diva2:328101/FULLTEXT02.pdf.

Reinhard, Erik, Erum Arif Khan, Ahmet Oguz Akyuz, and Garrett Johnson. *Color imaging: Fundamentals and applications*. CRC Press, 2008.

Sahabi, Hossein, and Anup Basu. "Analysis of error in depth perception with vergence and spatially varying sensing." *Computer Vision and Image Understanding* 63, no. 3 (1996): 447–461.

Chapter 4

Fundamentals of image analysis and interpretation

4.1 INTRODUCTION

This chapter describes how digital images are represented and investigates the basic concepts in the domain of image processing. Following this, some of the most popular image pre-processing algorithms are introduced. Pre-processing algorithms, such as denoising and contrast enhancement methods, are useful as they can improve the success and consistency of subsequent digital image analysis techniques. This chapter then looks at fundamental image-processing techniques that will be of particular interest to engineers and inspectors such as image alignment, edge detection, and morphological operations.

The final part of this chapter deals with camera calibration. In order for the camera to act as a measuring device, it must be geometrically calibrated to the physical world. Geometric camera calibration seeks to determine the relationship between pixel dimensions and real-world units, like millimeters, and also correct for lens distortion. This allows the image data to be expressed in physically meaningful units that can be readily interpreted by engineers. Sample examples are provided to help illustrate all of the concepts and techniques covered in this chapter.

4.2 IMAGE REPRESENTATION

In a numerical sense, an image is a discrete array of intensity samples that encodes visual information. The mathematical representation of an image as a function $f(x,y)$ is helpful for image processing tasks. $f(x,y)$ is the visual information such as color or brightness at position (x,y), where x and y are positive integer spatial indices in the horizontal and vertical directions, respectively, as shown in Figure 4.1. The function $f(x,y)$ is defined over a rectangular grid, with M elements in the horizontal direction and N elements in the vertical direction. The origins of the x and y direction are generally considered to be the top, left-most corner of the image.

$$
f = \begin{bmatrix} f(1,1) & f(1,2) & \cdots & f(1,M) \\ f(2,1) & f(2,2) & \cdots & f(2,M) \\ \vdots & \vdots & \ddots & \vdots \\ f(N,1) & f(N,2) & \cdots & f(N,M) \end{bmatrix}
$$

Figure 4.1 An image can be viewed as a matrix. Each element of the matrix is called a pixel.

Each element in this array is called a pixel (picture element). The image represented in Figure 4.1 consists of only one 2D array—the value of each pixel is proportional to the brightness or gray levels of the image at that point. An RGB (Red, Green, Blue) color image consists of three 2D arrays stacked together and can be written as a "vector-valued" function, as per Equation 4.1, where r, g, and b refer to the red, green, and blue channels, respectively.

$$
f(x,y) = \begin{bmatrix} r(x,y) \\ g(x,y) \\ b(x,y) \end{bmatrix} \tag{4.1}
$$

4.2.1 Image types and pixel bit-depth

The value of pixels in an image is defined up to a limited precision. One of the most common pixel formats used is 8-bit RGB. Eight bits can represent 256 different "strengths" of red, green, or blue, from 0 to 255. A value of 0 indicates none of that color is present, while a value of 255 indicates the color is present in full strength. Taken together, there is a possible palette of $256 \times 256 \times 256$ or 16.8 million colors.

The color of a pixel depends on the strengths of the component primary colors. When one of the components has the strongest intensity, the color is a hue near this primary color (reddish, greenish, or bluish), and when two components have the same strongest intensity, then the color is a hue of a secondary color (a shade of cyan, magenta, or yellow). A secondary color is formed by the sum of two primary colors of equal intensity: cyan is green plus blue, magenta is red plus blue, and yellow is red plus green. For example, pure red corresponds to a value of 255 in the red channel and 0 in the green and blue channels, while pixels that have a yellow hue have high values in the red and green channels and low values in the blue channel—something that may be observed in the color image in Figure 4.2 where the pixel values corresponding to the yellow-ish surface have high red and

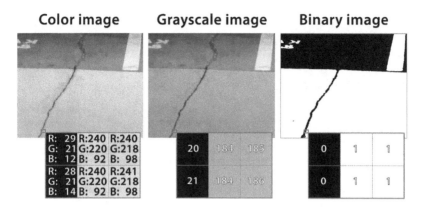

Figure 4.2 **Examples of various image types.**

green components and a relatively small blue component. The pixels that represent the crack are much darker than the surrounding yellow surface, and for this reason, the values across all color channels are small.

A grayscale image can be obtained from a color image by taking the average of each of the color channels (or some other linear combination of the channels). A grayscale image can be reduced to a binary image, which consists exclusively of 0s (black) and 1s (white), by specifying a threshold value and replacing all pixels in the grayscale image with intensities greater than the threshold level with the value 1 (white) and replacing all other pixels with the value 0 (black). In MATLAB®, the threshold level can be (i) specified by the user; (ii) computed based on Otsu's method, which uses the value that minimizes the intraclass variance of the black and white pixels; or (iii) the default threshold value can be used, which is midway between pure black and pure white. The following code shows how to read in and display images, and how to convert color images into grayscale and binary images in MATLAB®.

```
1  % Display Color, Grayscale and Binary Image
2  % Read in color image
3  Color_Image = imread('Crack.jpg');
4  figure, imshow(Color_Image), title('Original Color Image')
5
6  Gray_Image = rgb2gray(Color_Image);
7  figure, imshow(Gray_Image), title('Grayscale Image')
8
9  % Compute a global threshold (level) that minimizes the
10 % intraclass variance of the black and white pixels
11 level = graythresh(I)
12 BW_Image = im2bw(Gray_Image, level);
13 figure, imshow(BW_Image), title('Binary Image')
```

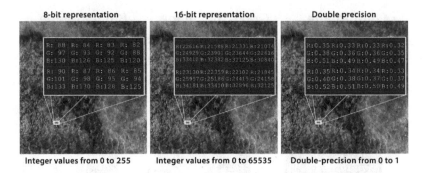

Figure 4.3 **The range of pixel values for common image representation formats.**

There are quite a few ways of representing color images other than using the 8-bit RGB format. For scenes with a wide dynamic range (images with detail in both the shadows and highlights), or for applications where capturing subtle color variations is important, the 256 discrete values offered by the 8-bit format may not provide enough precision. MATLAB® also supports 16-bit and floating-point double-precision formats (among others). The 16-bit format allows representing the color data in a scene with a greater degree of precision than 8-bit, as there are $2^{16} = 65,536$ discrete levels for each color channel instead of the $2^8 = 256$ levels available to 8-bit images. The floating-point double-precision format is useful as a working format because quantization and truncation errors are kept to a minimum, which is particularly important when performing certain image processing operations. For example, if an unsigned 8-bit image is subtracted from another unsigned 8-bit image, everywhere the result is a negative would be represented as a 0 as unsigned integers cannot represent negative numbers. Switching from integer representation to floating-point representation solves this problem. This can be done in MATLAB® using the function im2double(). It is worth noting that images must often be in the same format before being combined with one another through addition, subtraction, multiplication, or division. For this reason, it is generally a good idea to convert all images into the same format. The range of pixel values for each format is illustrated in Figure 4.3.

4.2.2 Color spaces

There are various ways that color can be represented. Humans may describe a color by its attributes of brightness, hue and, colorfulness, a printing press may produce a specific color based on the reflectance and absorbance of cyan, magenta, yellow, and black inks on the printing paper, while an LCD monitor or TV screen may render a color based on the intensity of the red, green, and blue subpixels at each pixel location. A color space, also known as a color model, is a specific organization of colors that allows color information to be represented numerically. Generally, it is observed that a color can be specified using three coordinates, or parameters. These parameters describe the position of the color within the color space and are sometimes referred to as the "tristimulus values." There is a vast array of color spaces; popular ones in the domain of image processing include RGB, HSV, and $L^*a^*b^*$. There are many more color spaces that are very similar to these color spaces.

The RGB color space is an additive color space in which red, green, and blue light are added together in various ways to reproduce a broad array of colors (Poynton, 2003). The RGB color space does have some shortcomings, however. It is psychologically non-intuitive as humans may find it difficult to visualize a color defined by red, green, and blue attributes. Another disadvantage with the RGB color space in applications with natural images is a high correlation between its components, as well as a perceptual non-uniformity, that is, the low correlation between the perceived differences of two colors. Perceptually non-uniformity means that a relatively large change in color value does not necessarily produce a noticeable change of color. The high correlation between components stems from the fact that red, green, and blue components in an RGB image are all correlated with the same amount of light hitting the object, and therefore with each other (Cheng et al., 2001). This issue is evident in Figure 4.4, which depicts each channel individually for the RGB, HSV, and $L^*a^*b^*$ versions of the same image. It may be observed that the red, green, and blue channels in the RGB color space are quite similar to one another. Descriptions in terms of

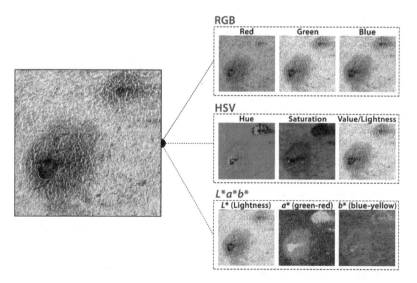

Figure 4.4 Visualization of each channel of a color image in the RGB, HSV, and $L^*a^*b^*$ color space.

hue-saturation-brightness are often more relevant due to this separation of chromatic and achromatic information.

HSV (hue, saturation, value) is one of several variations of color spaces characterized by the factors in the parenthesis. The hue corresponds to the predominant color, saturation indicates the "purity" of the color, and the value indicates the amount of light. The $L^*a^*b^*$ space consists of a luminosity layer L^* and chromaticity layers a^* and b^*. The L^* component closely matches the human perception of lightness. The color information is stored in the a^* and b^* layers. The a^* component indicates where the color lies on the red-green axis, while the b^* component indicates where the color lies on the blue-yellow axis. The $L^*a^*b^*$ space attempts to reflect a uniform change in perceived color with a corresponding uniform change in the L^*, a^*, and b^* components.

Color space conversion is the translation of the representation of a color from one basis to another. This typically occurs in the context of converting an image that is represented in one color space to another color space. Given how common color space conversions are, it is not surprising that MATLAB® has dedicated functions for converting between RGB and HSV: rgb2hsv() and hsv2rgb(), as well as between RGB and $L^*a^*b^*$: rgb2lab() and lab2rgb().

4.3 PRE-PROCESSING ALGORITHMS

Raw image data that directly comes from a camera is likely to have a number of problems such as dead pixels, geometric lens distortion, vignetting, uneven lighting across a scene, noise, and inaccurate color reproduction, among others. Naturally, these problems can diminish the performance of subsequent image analyzing algorithms. Pre-processing algorithms can be used to enhance image features that are important for further processing, attenuate insignificant features (such as noisy artifacts), or to correct for distortions (both photometric and geometric distortions). Some common pre-processing operations are presented in Figure 4.5

This section deals with:

- Point operations—these operations deal with one pixel at a time, with no consideration of neighboring pixels. Other point processing examples include color conversions and numeric data conversions.
- Neighborhood operations—these operations consider the local neighborhood around the processed pixel when computing the output. Examples include filtering-based operations and morphological operations.
- Image restoration/enhancement methods using multiple images—these operations rely on data from multiple images to create one "super" image that contains more information than any of the constituent images.

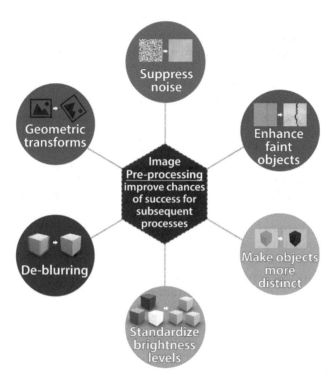

Figure 4.5 Image-pre-processing operations.

- Geometric transformations—these operations modify the spatial relationships between pixels in an image. They are often used for image alignment and correcting for lens distortion.

4.3.1 Point operations

For point operations, each pixel in the output image is a function of the intensity of the pixel at the corresponding position in the input image. Since point operations only consider the pixel value at a single point, independent of neighboring pixels, they are normally quick to compute. Common point operations include thresholding and contrast enhancement (pixel brightness transformation). In the context of point operations, it is important to discuss histograms.

Histograms

The histogram of an image is a graph showing the distribution of pixel intensities in that image. For an 8-bit grayscale image there are 256 different possible intensities, and so the histogram will graphically display 256 bars or points showing the distribution of pixels among those grayscale values. For color images, either individual histograms of red, green, and blue channels can be taken, or a 3D histogram can be produced, with the three axes representing the red, blue, and green channels (Ghosh et al., 2011).

Histograms have many uses. One of the more common is to decide what value of threshold to use when converting a grayscale image to a binary one by thresholding. A good way to choose a threshold value is to visually inspect the image histogram. If the image is well-suited for thresholding, then the histogram will be bi-modal, that is, the pixel intensities will be clustered together into two well-separated sections. Such a scenario is presented in Figure 4.6, where there is an image of a dark damaged region on a light surface. The histogram for this image has two dominant distinct modes, with one representing the light background pixels and the other representing pixels in the dark damaged area. One obvious way to separate the damaged area from the background is to select a threshold value somewhere in between the two peaks in the histogram. If the distribution is not like this, then it is unlikely that a good segmentation can be produced by thresholding.

Histograms can reveal a lot about the overall brightness and contrast in an image. Contrast is an important factor in any subjective evaluation of image quality. It is the difference in luminance or color that makes an object in an image distinguishable from its surroundings. It may be observed from the histogram in Figure 4.6 that the full range of gray values is being utilized. This is indicative of a high contrast image. However, for some images, the distribution of gray values is concentrated within a narrow band on the histogram, as is the case for the images shown in Figure 4.7. For the image in Figure 4.7(a), the peak of the histogram lies between 0 and 50 on the x-axis (pixel intensity scale), which reflects the fact that this image has a dark appearance. For the image in Figure 4.7(c), which is at

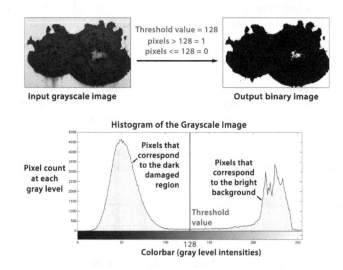

Figure 4.6 Using histograms to help choose a suitable threshold value.

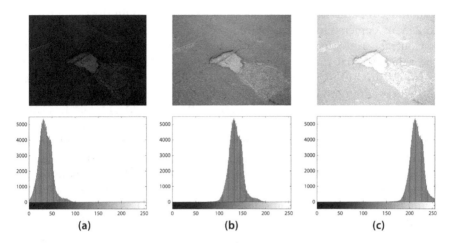

Figure 4.7 The histograms of (a) a dark, (b) a balanced image, and (c) a bright image.

the other end of the spectrum, it is clear from the histogram that the bulk of the pixels have intensity values greater than 200, which reveals that the overall image is very bright.

For each image, it is clear that the complete intensity range is not being utilized and the overall contrast is quite poor. Images such as these could benefit greatly from contrast enhancement.

Contrast enhancement

Contrast enhancement is the process of taking a low contrast image, such as the image shown in Figure 4.8(a) where the intensity values lie between

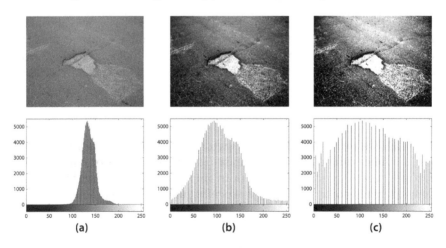

Figure 4.8 The histograms of the (a) original image, (b) contrast enhanced using the function imadjust(), and (c) contrast enhanced using the function histeq().

100 and 200, and "stretching" the gray level limits to fill the range from 0 to 255, as depicted in Figure 4.8(b).

Contrast stretching, in its simplest form, is achieved by first shifting all the values so that the gray level range begins at 0, that is, the new range is from 0 to 100 (this histogram would look like the histogram depicted in Figure 4.7(a)), and then multiplying all the intensity values by a factor of 2.55 such that the new range becomes 0 to 255. MATLAB® has some dedicated functions for performing contrast enhancement. The first one is imadjust(), which maps the intensity values in an input grayscale image A to values in a new image B such that 1% of the pixel intensity values are saturated at the lowest and highest intensities. The result of this is shown in Figure 4.8(b). The second function is histeq(), which enhances contrast using histogram equalisation (Finlayson et al., 2005). It operates by transforming the intensity values in input image A in such a way that the gray levels in the output image B are approximately evenly distributed across all discrete intensity levels—the effect of this will be that the histogram of B will be flatter than the histogram of image A. The result of using this function is shown in Figure 4.8(c). An adaption to this global histogram equalization is a local-based operation called contrast-limited adaptive histogram equalization (Zuiderveld, 1994), which can be performed using the MATLAB® function adapthisteq(). This function operates on small regions in the image and it has the effect of enhancing local contrast.

While contrast enhancement can help to improve the clarity of images, care should be taken to avoid excessive contrast stretching, which can amplify any noise present in the original imagery and cause issues such as color banding and give the false appearance of harsh surface textures.

4.3.2 Neighborhood operations

Unlike point operations, which work on a per-pixel basis, neighborhood operations consider the local neighborhood in the input image around the chosen pixel for computing the output. This section looks at filtering methods, in particular, denoising and edge enhancing filters. Following this, morphological operations are covered.

Filtering

Filtering is a technique for modifying or enhancing an image. For example, an image can be filtered to emphasize certain features or remove other features. Image-processing operations implemented with filtering include smoothing for noise reduction, sharpening, and edge enhancement.

De-noising

Filtering-based de-noising methods exploit the redundant visual information in images. Neighboring pixels corresponding to a single object in real images will usually have essentially the same or similar brightness values, so if a "noisy" pixel can be identified from the image, it can usually be restored as an average value of neighboring pixels. This is the basis for noise reduction filters. One of the most common assumptions made about noise is that it has a high spatial frequency. In this case, it is often adequate to apply a low-pass spatial filter, which will attenuate the higher spatial frequencies while allowing low spatial frequency components to pass through to the output image. Of course, if the original image itself exhibits high spatial frequencies, such as an image of a richly textured surface, then it will be somewhat degraded after filtering.

These low-pass filters can be implemented by convolving the image with a mask. The values in the mask can be viewed as weighting factors. The value of each output pixel is found by centering the mask over the corresponding pixel in the input image, and computing the sum of the pixel intensity values at and around the corresponding point weighted by the mask values depending on the relationship with the center pixel. For example, each of the mask values might be equally weighted, in which case the operation is simply the evaluation of the local mean of the image in the vicinity of the mask. This type of filter is often called an averaging filter. An example of this is shown in Figure 4.9 where a 3-by-3 filter containing equal weights (that sum to 1 so there is no net change in the overall brightness of the image) visits each pixel in the input image.

With reference to Figure 4.9, it may be noted that the effect of the filtering is that the value of the pixel in the output image is the average of the pixels

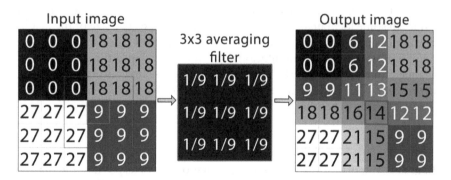

Sample computation of output pixel
Output pixel=0(1/9)+18(1/9)+18(1/9)+27(1/9)+9(1/9)+9(1/9)+27(1/9)+9(1/9)+9(1/9)
Output pixel=0+2+2+3+1+1+3+1+1
Output pixel=**14**

Figure 4.9 **A 3-by-3 averaging filter and demonstration of the mask at a point in the** image.

in the 3-by-3 neighborhood around the corresponding point in the input image. This averaging, or smoothing effect, is applied at all points in the image. It can be performed in MATLAB® using the function imfilter(), as seen below:

```
1   % Read in image
2   Color_Image = imread('Crack.jpg');
3
4   % Create a 3-by-3 filter having values of 1/9
5   h = ones(3,3)/9;
6   % Aside: Using the function fspecial() can be useful to get
7   predefined filters. E.g. "h = fspecial('gaussian', [3,3], 1);"
8   creates a 3-by-3 Gaussian filter standard deviation of 1.
9
10  % Apply filter to image. Use optional tag 'replicate' to pad
11  the image with pixels with the same value as adjacent
    pixels
12  at the image boundary for better results near the
    boundary.
13  Output_Image = imfilter(Color_Image, h, 'replicate');
```

Occasionally, it may be more useful to apply this smoothing subject to some condition, for example, the center pixel is only assigned a new value if the difference between the average value and the original pixel value is greater than some predefined threshold. This goes some way toward removing noise while preserving real edges and detail in the original image.

Another common noise suppression technique is median filtering, where a pixel in the output image is assigned the value of the median of pixel values in some local neighborhood in the input image. The size of the neighborhood is arbitrary, but neighborhoods in excess of 3×3 or 5×5 may be impractical from a computational point of view since the evaluation of the median requires that the image pixel values be first sorted. In general, the median filter is superior to the mean filter in that noise is reduced while image blurring is kept to a minimum.

Edge enhancement and detection

Edges are important features of images that often correspond to the outline of real objects in a scene. Therefore, highlighting and identifying the edges in an image is useful in isolating objects of interest and gaining an improved understanding of the scene. One of the most popular edge-emphasizing filters is the Sobel operator (Sobel, 1990). The Sobel operator is a two-dimensional filter that calculates approximations of the first derivatives of an image in the horizontal and vertical directions. The magnitude of the

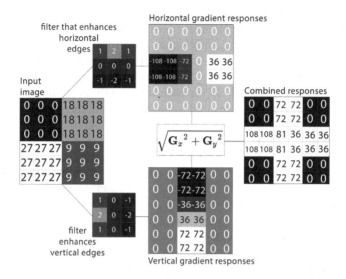

Figure 4.10 Demonstration of the Sobel edge detector.

output of the Sobel operator will be large at places in the image where there is a sharp change in the intensity values between neighboring pixels, which is often indicative of an edge. This is illustrated in Figure 4.10, where the Sobel filter is applied to approximate the horizontal gradient responses of the input image. The transpose of the filter is used to perform a spatial-derivative operation for enhancement of vertical edges.

Denoting G_x and G_y as the two masks that give the horizontal and vertical derivative approximations at each point as:

$$
G_x = \begin{bmatrix} 1 & 2 & 1 \\ 0 & 0 & 0 \\ -1 & -2 & -1 \end{bmatrix} * A \text{ and } G_y = \begin{bmatrix} 1 & 0 & -1 \\ 2 & 0 & -2 \\ 1 & 0 & -1 \end{bmatrix} * A \tag{4.2}
$$

where the asterisk denotes the 2-dimensional convolution operation. A padding with a thickness of one pixel may be applied around the border of image A during the convolution process, thereby enabling the computation to be performed at the image periphery. The padded pixels can assume the value of the neighboring pixels in the original image. At each point in the image, the resulting gradient approximations can be combined to give the gradient magnitude, using:

$$
G_{total} = \sqrt{G_x^2 + G_y^2} \tag{4.3}
$$

Large values in this image, G_{total}, represent sharp changes in image intensity, which in turn is indicative of an edge boundary. A common subsequent step is to threshold this image such that only the strong edges are represented. This entails assigning a value of 1 to all pixels where the gradient magnitude is greater than some predefined threshold, and 0 everywhere else. Finding the best threshold for each image will normally involve some trial and error. A lower threshold value will mean that smaller gradient magnitudes will be considered as edges, while a higher value will restrict the number of pixels that are considered as edges to only the larger gradient magnitudes.

Another common method for finding edges is the Canny method (Canny, 1987), which is distinct from other edge-detection methods in that it has the ability to detect both strong and weak edges. This is achieved by using two different thresholds (one threshold sensitive to strong, distinct edges, and another threshold that is effective in detecting weaker, more subtle edges). The Canny edge detector works initially by using a Gaussian filter to limit the effect of noise in the imagery. The intensity gradient of the Gaussian filtered image is then determined and "non-maximum suppression" is employed to ensure that the detected edges are as close to the actual object boundary as possible by searching for local maxima.

MATLAB® has a few inbuilt functions that make computing the image gradients and finding edges somewhat easier. These functions are presented below:

```
1   % Read in image and convert to grayscale
2   Color_Image = imread('Crack.jpg');
3   Gray_image = rgb2gray(Color_Image);
4   % Find the gradient magnitude, Gmag, and the gradient
5   direction, Gdir
6   [Gmag,Gdir] = imgradient(Gray_Image);
7
8   % Find edges in a grayscale image using Sobel operator.
9   % The Sobel operator is used by default. The threshold
    is a value between 0 and 1. If omitted, it is chosen
    automatically
10  Edges_Sobel = edge(Gray_image,'Sobel', 0.5);
11
12  % Alternatively, the Canny method can be used to find
    edges.
13  Edges_Canny = edge(Gray_image,'Canny');
```

Application of the edge detection method using the Sobel operator is shown in Figure 4.11.

Edge-enhancing filters and edge detection methods are frequently used for subsequent image-processing application such as contour matching or

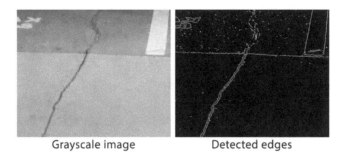

Grayscale image   Detected edges

Figure 4.11 Demonstration of the Sobel edge detector on an input grayscale image. The white pixels in the binary image represent places where there is a sharp change in pixel intensity values—something that often occurs at edges of real-world objects.

object detection. They are also widely used for real-time applications, such as traffic monitoring tasks or real-time stereo matching, where most of the image data is discarded and only the edges in a scene are used as this can facilitate faster processing times.

Morphological operations

Morphology deals with filtering based on geometric properties. The most fundamental morphological operators are dilation and erosion. Dilation makes an object larger by adding pixels to its boundaries, while erosion has the opposite effect by removing pixels from object boundaries. This is demonstrated in Figure 4.12, where the goal is to isolate the contiguous region corresponding to the worn-away bridge surface. Performing a simple thresholding operation provides a good starting point; however, it may be noticed that there are a lot of spurious white pixels remaining in Figure 4.12(b). These spurious pixels are dotted throughout the image in small clusters. Most of these spurious pixels can be removed by performing an erosion operation, as shown in Figure 4.12(c). Erosion is a process by which a structuring element (can be any shape with a specified size, e.g., a circle of radius 15 pixels) works its way throughout the image and removes pixels from the boundaries of objects (the number of pixels removed depends on the structuring element size). The erosion operation has the effect of removing objects that cannot completely contain the structuring element, for instance, if a circular object of radius 14 pixels was present in the image, it would be completely removed by the structuring element of radius 15 pixels.

Following erosion, the size of the remaining regions can be restored by performing a dilation operation using the same structuring element as used for the erosion operation, as shown in Figure 4.12(d). This operation adds pixels to the boundary of objects.

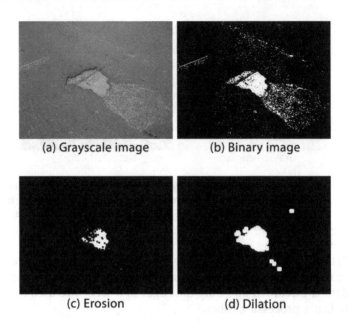

(a) Grayscale image   (b) Binary image

(c) Erosion   (d) Dilation

Figure 4.12 (a–d) Morphological operations.

Erosion and dilation are often used together. When an image is eroded and then dilated, the combined operation is called "opening." This is good for removing small, insignificant regions without overly shrinking large regions. When an image is dilated and then eroded, the combined operation is called "closing." This operation is good for merging small isolated regions that are close together into one contiguous large region.

A related concept involves "filling-in" holes in a region. Holes are considered to be an "island" of black pixels completely surrounded by white pixels in a binary image. This operation is particularly useful when one is aware of the nature of the detected regions. For example, if it is known that instances of damage occur in large contiguous "blobs," then a safe assumption to make would be that all pixels within a detected region, including small clusters of pixels that are classified as being non-damaged, are most likely a part of the damaged region. The most common morphological operations in MATLAB® can be carried out with the following commands:

```
1  % Read in image
2  Im = imread('exposed bridge deck.tif');
3
4  % Convert to black and white and display image
5  BW = im2bw(Im,0.6); imshow(BW)
6
```

```
7 % Define circular structuring element of radius = 5
8 SE = strel('disk',5);
9
10 % Perform erosion and see the result
11 ErodedBW = imerode(BW,SE); figure, imshow(ErodedBW)
12
13 % Perform dilation on the eroded image
14 DilatedBW = imdilate(ErodedBW,SE); figure, imshow(DilatedBW)
15
16 % Fill in white regions and view the result
17 FilledBW = imfill(BW,'holes'); figure, imshow(FilledBW)
```

A morphological approach to segmentation is particularly attractive because it produces closed, well-defined regions, and provides a framework in which prior knowledge about the images in a particular application can be utilized to improve detection results.

4.3.3 Image restoration/enhancement methods using multiple images

For underwater inspections, imagery is almost always captured using a handheld camera, which is difficult, if not impossible, to keep completely static. For top-side inspections, however, cameras can be mounted on a tripod, which opens up the possibility of capturing multiple images of the exact same scene (assuming the scene remains static). By merging data from multiple images, a single image can be produced that contains more information about the scene than any of the constituent images. An example of this is high dynamic range (HDR) images, which are formed by combining two or more standard dynamic ranges images that are captured at different exposures—thereby allowing data in both the dark and bright parts of a scene to be represented in a single image. While some cameras provide a dedicated HDR mode that automatically merges several standard dynamic range (SDR) photographs, other cameras do not offer this feature or can only slightly extend the dynamic range. This section looks at how HDR images can be created manually. A denoising technique is also discussed—it involves averaging data from multiple images of the same scene so as to create a composite image with less noise than the constituent images. This technique can be particularly useful in dark scenes (which require the camera to use high ISO values that amplify the level of noise in an image).

Generating HDR imagery

As discussed in Chapter 3, HDR involves merging two or more photographs of the same scene, taken at various exposure levels, into one completed image. One dark exposure can capture detail in bright areas of the scene, while a lighter exposure can give details in mid-tones and shadows. Combined, the standard images can give a finished image with far more overall tonal detail than any of the individual SDR images. This is especially applicable for low-light and backlit scenes where HDR can brighten up the dark parts of the scene without washing out the well-lit portions of the scene.

The process for capturing and creating HDR images can be summarized as follows:

1. Capturing bracketed exposures
2. Generating the HDR image
3. Tone mapping of the HDR image

Capturing bracketed exposures

Many current cameras on the market support auto-exposure bracketing, whereby the camera automatically takes a number of photographs of the scene at different exposure settings. If this feature is not available, the user can capture several SDR photographs using different shutter speeds (while keeping the other parameters constant). The faster the shutter speed, the less light reaches the sensor, and hence the image is darker (underexposed). Underexposed images preserve details in the bright areas of a scene. Conversely, if the shutter is open for a longer period of time, then more light reaches the sensor and the resulting image is overexposed. Overexposed images capture details in the dark areas of a scene. Generally speaking, three photographs (underexposed, normally exposed, and overexposed) are sufficient; however, more photographs will tend to produce better results.

Generating the HDR image and tone mapping

Combining images can be done using various merging algorithms such as the one developed by Debevec and Malik (1997). Their algorithm relied on knowing the response function of the digital camera, which is the relationship between the physical light intensity reaching a pixel (on the sensor of the camera) and the intensity that appears in the image. With the response function, multiple photographs could then be merged into a single, high dynamic range radiance map, with the pixel values being proportional to the true radiance values in the scene. The code to create HDR images in MATLAB® is shown below. Since HDR images cannot be properly displayed on a display with limited dynamic range, tone mapping is employed to reduce the dynamic range, or contrast ratio, of the entire image, while retaining localized contrast.

```
1  % Specify filenames of the constituent SDR images
2  files = {'UnderExp.jpg', 'NormalExp.jpg', 'OverExp.jpg'};
3
4  % Specify shutter speed for each SDR image (Check EXIF info)
   expTimes = [1/13, 1/3, 1];
5
6  % Create HDR image
7  hdr = makehdr(files,'RelativeExposure',expTimes./
   expTimes(1));
8
9  % Tone map image to view on a display with limited
   dynamic range rgb = tonemap(hdr);
10
11 % Display tonemapped HDR image
12 figure; imshow(rgb)
```

The input SDR images and output HDR image corresponding to the above code is shown in Figure 4.13, which features an aging wharf pile in a marine environment.

The main challenge of merging images is any sort of movement, whether it be moving objects in a scene or movement of the camera. That is why exposure bracketing is most often applied to static objects and the camera is mounted on a tripod to ensure perfect alignment. This obviously imposes limitations on what can be photographed using HDR. When working freehand, there is a likelihood that all the captured images will not be perfectly

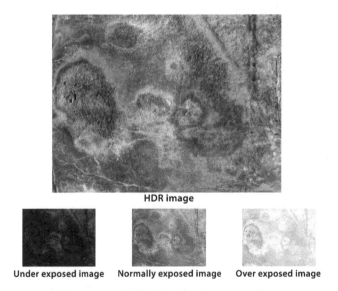

HDR image

Under exposed image   Normally exposed image   Over exposed image

Figure 4.13 HDR image that was created using three SDR images.

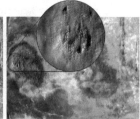

| HDR image with aligned SDR images | HDR image with improperly aligned SDR images | Alignment errors that must be corrected |

Figure 4.14 HDR images can have unwanted artifacts if the SDR images are not correctly aligned.

aligned. If this is the case then the resulting HDR image will appear fuzzy, as illustrated in Figure 4.14.

Image alignment

MATLAB® can perform image registration, which is very effective in correcting the misalignment. Image registration works by determining the parameters of the spatial transformation needed to bring the images into alignment. The spatial transformation may include shifting, rotating, warping, and stretching the images to ensure alignment. A sample code for registering two images is presented below. This code can readily be extended to register multiple images.

```
1 % Read in two images to be aligned
2 OE  = imread('OverExp.jpg');
3 UE = imread('UnderExp.jpg');
4
5 % Detect SURF features in each image
6 points1 = detectSURFFeatures(rgb2gray(OE));
7 points2 = detectSURFFeatures(rgb2gray(UE));
8 [f1,vpts1] = extractFeatures(rgb2gray(OE),points1);
9 [f2,vpts2] = extractFeatures(rgb2gray(UE),points2);
10
11 % Match features
12 indexPairs = matchFeatures(f1,f2);
13 matchedPnts1 = vpts1(indexPairs(:,1));
14 matchedPnts2 = vpts2(indexPairs(:,2));
15
16 % Estimate the transformation
17 [tform,inlierPtsDistorted,inlierPtsOriginal] =
18 estimateGeometricTransform(matchedPnts2,matchedPnts1,'p
   rojective')
```

```
19
20 outputView = imref2d(size(OE));
21 Registered_UE = imwarp(UE,tform,'OutputView',outputView);
22 figure; imshow(Registered_UE); title('Recovered image');
```

Advances in cameras such as the introduction of high-speed bracketing reduce the limitations imposed on HDR and even enable good quality photos to be captured freehand (albeit with a steady hand), thereby extending HDR imagery beyond its widely underutilized role as just an artistic technique, to new and practical purposes.

Noise suppression by image averaging

The rationale behind noise suppression by image averaging is quite simple: in the averaging process, the constant part of the image (that which is due to light reflected from stationary objects) will remain unchanged while the noise will be essentially random and will differ in each image. The assumptions inherent in this approach are as follows:

1. The noise in each image is random and independent of the noise in other images.
2. The noise has a zero-mean value.

With these assumptions, it is possible to show that averaging images increases the signal-to-noise ratio by an amount that is proportional to the number of images that are averaged. A demonstration of image averaging is presented in Figure 4.15. The value of each pixel in the output image is found by taking the average of the pixel values at the corresponding location in the input images.

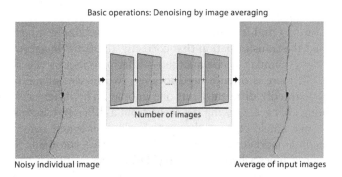

Figure 4.15 Noise reduction by averaging several images.

It may be observed in Figure 4.15 that the noise in the output image is greatly reduced compared to each of the input images. A quick example of how such an operation can be performed in MATLAB® is as follows:

```
1 % Read in 4 images and compute their average.
2 Im1  = imread('Image1.jpg');
3 Im2  = imread('Image2.jpg');
4 Im3  = imread('Image3.jpg');
5 Im4  = imread('Image4.jpg');
6 % Take the average of the input images
7 Averaged_Image = 1/4*Im1 + 1/4*Im2 + 1/4*Im3+ 1/4*Im4;
```

Although image averaging is shown here, the same basic statistical principle can be applied to similar types of operations, such as finding the median value at each pixel location, which also serves to attenuate noise. Additionally, image subtraction finds many uses in image processing. It can be useful for the subtraction of a known pattern (or image) of superimposed noise or, indeed, for motion detection: stationary objects cancel each other out while moving objects are highlighted when two images of the same dynamic scene, which have been taken at slightly different times, are subtracted. This process of subtraction of an uninteresting background image from a foreground image containing information of interest is referred to as "background subtraction."

Background subtraction is also useful when performing photometric calibration or when accounting for non-uniform lighting across a surface. By taking an image of a uniformly colored object, such as a sheet of white paper that is placed on top of the surface of interest, the variations in lighting can be identified. Images of the surface that are subsequently acquired can then be processed by subtracting this calibration image from them.

4.3.4 Geometric transformations

The previous subsection showed the importance of image alignment when creating HDR images from several input images. In such a case, it was quite straightforward to find matching points between each input image and subsequently find the transformation that correctly aligns the images because all images were very similar to one another—differing only in terms their exposure. However, there is often a need to align two images of the same scene that appear markedly different from one another. The differences can arise due to lighting changes, viewing the scene from a different perspective, using a different camera/field of view, the scene having undergone some evolution, or due to a combination of these factors. In such cases, automatic registration using the approach described for aligning SDR images is unlikely to work well. Instead, it will be necessary to identify several corresponding points in the two images and use these as the control points when generating the transformation

(a)   (b)

Figure 4.16 Original photographs taken at inspections in (a) 1987 and (b) 1996. (From: National Bridge Inspection Program, Vermont Agency of Transportation. With permission.)

that aligns the images as closely as possible. An example of this is presented in Figure 4.16, which features inspection photographs of a bridge taken several years apart. This example illustrates one of the strengths of using image analysis as part of the inspection regime; by comparing images taken of the same regions at different times, the rate of propagation of damage may be calculated.

```
1  %Register two images using control points
2  Base_photo = imread('1987.tif'); figure,
   imshow(Base_photo)
3  unregistered = imread('1996.tif'); figure,
4  imshow(unregistered)
5
6  [movingPts,fixedPts] =
7  cpselect(unregistered,Base_photo,'Wait',true);
8
9  mytform = fitgeotrans(movingPts, fixedPts, 'projective');
10
11 registered = imwarp(unregistered,
12 mytform,'OutputView',imref2d(size(orthophoto)));
13 Z = imlincomb(0.6,registered,0.4,orthophoto);
14 imshow(Z)
```

The MATLAB® code for registering these two images is shown below.

The command `cpselect()` will bring up an interactive window where the user will be invited to select corresponding points in both images. For best results, as many points should be chosen as possible and users should aim to select corresponding points that are well-dispersed throughout the image—this will help minimize errors in the resulting transformation. The results of image registration using this approach is shown in Figure 4.17.

Geometric transformations are also used to correct lens distortions in images. The transformations are found through a process known as camera calibration, which will be discussed in the next section.

Figure 4.17 **Result of image registration. (Adapted from: National Bridge Inspection Program, Vermont Agency of Transportation. With permission.)**

4.4 CAMERA CALIBRATION

Lenses involve a certain degree of distortion and the image is often non-linearly distorted (see Section 3.2.2 for more details on lenses). While it is always best to use lenses that are as distortion-free or distortion-corrected as possible, software-based approaches can considerably help to correct for distortions; however, such software-based approaches are somewhat limited insofar as they cannot account for the spatial depth of objects in the scene. The effect of distortion is such that perfectly straight lines in the real world will appear to be slightly curved when imaged. Given this distortion, it is often difficult to extract precise measurements from images unless the camera has been geometrically calibrated.

Geometric camera calibration is the process of estimating the parameters of a lens and image sensor of an image or video camera. The parameters include camera intrinsic, distortion coefficients, and camera extrinsic. The extrinsic parameters define the position and orientation of a camera in 3D world space. The intrinsic parameters include the focal length, the optical center, also known as the principal point, and the skew coefficient. The distortion coefficients model radial and tangential distortion. Radial distortion occurs when light rays bend more near the edges of a lens than they do at its optical center. These radial distortions can usually be classified as either barrel distortions or pincushion distortions. In barrel distortion, the apparent effect is that of an image that "bulges" at the center, while for pincushion distortion, the visible effect is that the image center is compressed relative to the periphery of the image (Figure 4.18). Barrel distortion may be found in wide-angle lenses and is often seen at the wide-angle end of

Figure 4.18 Pin cushion and barrel distortion.

zoom lenses, while pincushion distortion is often seen in older or low-end telephoto lenses. Tangential distortion occurs when the lens and the image plane are not parallel, and mostly affects low-end lenses. Correcting for distortion is far more important if a wide-angle lens or cheap optics are used.

This section looks at calibrating a single camera for the purpose of removing lens distortion effects from an image and subsequently measuring the size of a defect in the scene applying some of the concepts discussed earlier in this chapter. The calibration of multi-camera systems is presented in Chapter 7. The process of undistorting images and measuring the size of observed defects is shown in Figure 4.19.

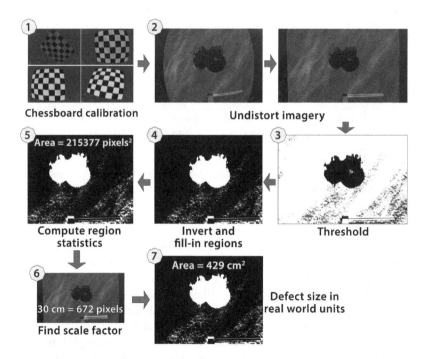

Figure 4.19 Process of undistorting imagery and measuring the size of observed defects.

Nowadays, cameras are most often calibrated offline using a special calibration pattern. A planar target with a checkerboard or circular pattern of known size is a well-established and popular method for camera calibration (Remondino & Fraser, 2006). Indeed, MATLAB® includes functionality for automatically detecting the corner points of a checkerboard and, after processing multiple images of the calibration pattern, can solve for the camera parameters.

The first step in checkerboard-based calibration involves capturing several images of a flat checkerboard (Figure 4.19(a)). It is important that the photographs of the calibration pattern are captured under the same conditions as the rest of the inspection imagery (i.e., the focal length and image resolution should be kept constant, the environment should remain the same—some of the camera parameters that apply for top-side scenes will be slightly different for underwater scenes). For best results, the calibration pattern should be photographed from a wide range of positions, ensuring that the whole of the calibration pattern is visible in the image, and as many photographs as possible should be captured, although 10 photographs are normally sufficient. The MATLAB® code for reading in calibration images, detecting the checkerboard, and estimating the camera parameters is shown below.

```
1  %Identify calibration files. Four images are used here
2  CalibImages = {'im01.jpg','im02.jpg','im03.jpg','im04.jpg'};
3
4  % Detect calibration pattern in the input images.
5  [imagePts, boardSize] = detectCheckerboardPoints(CalibImages);
6
7  % Generate world coordinates of the corners of the squares.
8  squareSize = 20;% The size of each square in millimeters
9  worldPts = generateCheckerboardPoints(boardSize, squareSize);
10
11
12 % Estimate the intrinsic, extrinsic, distortion parameters
13 [params] = estimateCameraParameters(imagePts, worldPts);
```

Once the calibration parameters have been found, these parameters can be applied to new images to correct for distortion, as shown in Figure 4.19(b). The following code reads in an image of a damaged region, undistorts the image using these recently found calibration parameters, and prints the area of damaged regions (that are greater than 100,000 pixels^2) on top of the black and white image, as shown in Figure 4.19(e).

In order for these dimensions to be physically meaningful for engineers, it is necessary to convert from pixel units to real-world units such as

```
1 % Remove lens distortion and display the results
2 I = imread('Damage.jpg');
3 I2 = undistortImage(I,params);
4 imshowpair(I,I2,'montage');title('Original & Corrected Image')
5
6 BW = im2bw(I2,0.3);% Threshold image (Fig. 4.19(c))
7 BW = imfill(1-BW,'holes');%Invert & fill holes (Fig. 4.19(d))
8
9 % Compute properties of white regions
10 CC = bwconncomp(BW);
11 s = regionprops(CC,'Centroid','Area');
12 figure, imshow(BW), hold on
13 % Overlay the area of each white region on the binary image
14 for i = 1:CC.NumObjects% Iterate through each white region
15    if s(i).Area >100000% Only consider large regions
16 text(s(i).Centroid(1),s(i).Centroid(2),['Area = ',...
17 num2str(s(i).Area),' pixels^{2}'],'Color','red','FontSize',45)
18    end
19 end
20 hold off
21
```

centimeters. This can be achieved by placing an object of known dimensions in the same plane as the defect of interest. In the above example, a ruler of length 30 cm was positioned close to the defect. The length of the 30 cm ruler corresponds to a length of 672 pixels in the image, which could be interactively measured using the function imtool()in MATLAB®. A scale factor could then be established with knowledge of this relationship. It is important to note that this scale factor only holds for the one plane that passes through the object of known dimensions (i.e., the ruler) and that is orientated normal to the camera. Using this scale factor, the size of detected regions can be expressed in terms of cm^2 in MATLAB as follows:

```
1 % Overlay the area of each white region on the binary image
2 sf = 30/672%Scale factor established using imtool()
3 figure, imshow(BW)
4 hold on
5 for i = 1:CC.NumObjects
6    if s(i).Area >100000% Iterate through each white region
7 text(s(i).Centroid(1),s(i).Centroid(2),['Area = ',...
8 num2str(s(i).Area*(sf^2)),'cm2'],'Color','red','FontSize',50)
9    end
10 end
11 hold off
```

The result of this operation is displayed in Figure 4.19(g).

4.5 SUMMARY

A key aim of this book is to outline the essentials that are of particular relevance to engineers and to provide sufficient material for readers to get up and running with developing their own algorithms, especially relating to damage assessment applications. This chapter provides an overview of the basic concept in the domain of image-processing and presents some fundamental image-processing operations and techniques that are often used as part of more sophisticated techniques. Following this, popular image pre-processing algorithms are covered. Image pre-processing algorithms are regularly used as, in practice, image analysis techniques rarely perform optimally when applied directly to photographs in their "as shot" state. Applying pre-processing algorithms to the images beforehand can help improve the success and consistency of subsequent digital image analysis techniques. An example of where pre-processing algorithms can be used effectively would be denoising a grainy image of a cracked surface. Performing noise reduction as an initial step will help to reduce the number of noisy artifacts that are erroneously detected by a crack detection technique. Some useful pre-processing algorithms, such as noise reduction and contrast enhancement, are described. These algorithms are demonstrated using simple examples coded in MATLAB®.

The final part of this chapter looks at camera calibration, which is an important part of the imaging pipeline and is something that engineers and inspectors will inevitably encounter. Camera calibration serves to connect pixel units in the image space to physically meaningful units in the real-world space, such as millimeters. It is used for a wide range of application such as measuring objects and distances, navigation systems, and for 3D scene reconstruction. As with the other concepts and techniques covered in this chapter, the calibration process is described in a step-by-step manner.

REFERENCES

Canny, John. "A computational approach to edge detection." In *Readings in Computer Vision*, pp. 184–203. 1987.

Cheng, Heng-Da, X. H. Jiang, Ying Sun, and Jingli Wang. "Color image segmentation: Advances and prospects." *Pattern Recognition* 34, no. 12 (2001): 2259–2281.

Debevec, Paul E., and Jitendra Malik. "Recovering high dynamic range radiance maps from photographs." In *Proceedings of the 24th annual conference on Computer graphics and interactive techniques*, pp. 369–378. ACM Press/ Addison-Wesley Publishing Co., 1997.

Finlayson, Graham, Steven Hordley, Gerald Schaefer, and Gui Yun Tian. "Illuminant and device invariant colour using histogram equalisation." *Pattern Recognition* 38, no. 2 (2005): 179–190.

Ghosh, Bidisha, Vikram Pakrashi, and Franck Schoefs. "High dynamic range image processing for non-destructive-testing." *European Journal of Environmental and Civil Engineering* 15, no. 7 (2011): 1085–1096.

Poynton, Charles. "Digital video and HDTV algorithms and interfaces." *Morgan Kaufmann Publishers, San Francisco* (2003): 260.

Remondino, Fabio, and Clive Fraser. "Digital camera calibration methods: Considerations and comparisons." *International Archives of Photogrammetry, Remote Sensing and Spatial Information Sciences* 36, no. 5 (2006): 266–272.

Sobel, Irvin. "An isotropic 3 × 3 image gradient operator." *Machine Vision for Three-Dimensional Scenes* (1990): 376–379.

Zuiderveld, Karel. "Contrast limited adaptive histogram equalization." *Graphics Gems* (1994): 474–485.

Chapter 5

Crack detection

5.1 INTRODUCTION

Visible cracks provide an indication of the structural degradation and are a key factor when diagnosing the condition of a structure. The presence of cracks can be unsightly, may cause a loss in serviceability, or, in more serious cases, can even lead to structural failure. Traditional crack assessments are often carried out as part of a visual inspection where observed cracks are mapped, counted, quantitatively measured, and photographed. This process is often costly and dull when performed by a human inspector. Even with great diligence, measuring the true extent of cracks is fraught with subjectivity and prone to error. Nevertheless, this task is crucial for ensuring the continued safety of structures.

Attempts to automate the crack detection process have received a lot of attention given the pervasive nature of cracks coupled with the tedious nature of manual crack assessments. This has led to the development of a number of image-based techniques such as wavelets (Khanfar et al., 2003), neural network approaches (Choudhary & Dey, 2012), statistical filters (Sinha & Fieguth, 2006), deep learning-based approaches (Cha et al., 2017) and percolation-based methods (Yamaguchi & Hashimoto, 2008). Wang and Huang (2010) conducted a comparison of crack detection methods and identified percolation-based methods as being particularly suited to detecting ambiguous cracks as well as being quite resilient to noise. However, they also recognized that conventional percolation-based methods are computationally expensive given that they operate on a local basis at each pixel location in the image. With this in mind, a modified percolation-based method is described in this chapter that offers increased efficiency over the classical percolation approach.

Understanding how techniques fare under different conditions is of high practical importance for engineers/inspectors. For this reason, the performance of the crack detection technique is investigated under a range of underwater visibility conditions. The technique is also applied to above-water scenes, where a number of other challenges are encountered, such as bright spots, difficult viewing angles, road markings/tire marks, and low-contrast monochromatic scenes, all of which may mislead the crack detection algorithm. This chapter concludes by looking at some good practices to follow when capturing photographs of cracks with a view to extracting physically meaningful information, such as crack widths/lengths in millimeters.

5.2 CRACK DETECTION TECHNIQUE

Cracks are generally characterized by their narrow shape and their darker appearance in comparison to the surroundings. Percolation-based methods take both of these characteristics into account when identifying cracks. Classical percolation-based methods operate by sliding a window over the image and analyzing the window contents at every position to assess whether a crack is present. The center of the window is treated as the seed point of the percolating region, which grows outward for several iterations. At each iteration, dark pixels that border the percolating region are incorporated into the percolating region. After the final iteration, the resulting pattern of dark pixels is analyzed. Cases where a narrow or linear pattern have been traced out are indicative of a crack, while irregular or radial diffusion patterns typically correspond to the non-cracked background.

While this process works well for detecting cracks, it is extremely slow as the window must visit every point in the image. This can be prohibitively slow for high-resolution images. To drastically speed up the process without harming detection accuracy, we can specify that the window should only visit locations in the image where an edge has been already identified since cracks will almost always produce an edge response. Computationally inexpensive edge detectors such as the Sobel operator or the Canny edge detector (which are discussed in Section 4.3.2) can be used to identify edges in an image. Of course, many of the edges will not correspond to a cracked region but rather will represent the boundary between two objects in the scene. A good way to prevent these non-crack edges from being classified as a crack is to impose a constraint that stipulates both sides of the crack should have a similar color composition, which would be expected for cracks dissecting a homogenous material. An overview of the methodology of the modified percolation-based technique is shown in Figure 5.1. This is followed by a step-by-step description of the technique.

The proposed method is described in a step-by-step manner as follows:

Step 1: The input is a color or greyscale image, A. The Sobel operator is applied to this image, which gives an approximation of the image gradient. The output from this operation is a binary image, where white pixels represent locations that have a large gradient. In the case of color images, the average gradient magnitude from the three color channels is used.

Step 2: This binary image identifies points in the image where performing percolation would be relevant while overlooking points in the

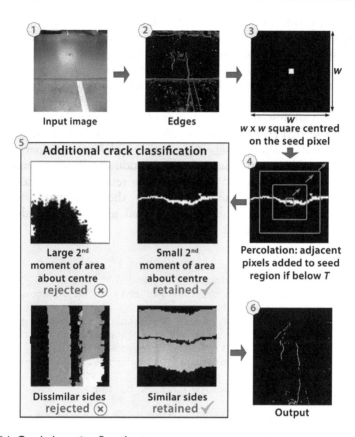

Figure 5.1 Crack detection flowchart.

image where cracks are unlikely to be present. This cuts down on the computation time considerably. It is not essential that the entire crack is detected with the Sobel operator. Partial detection should be sufficient as the window will likely overlap with non-detected portions of the crack. The detected pixels from this step are considered for percolation. The first two steps can be carried out in MATLAB® as follows:

```
1 % Read in color image
2 Image = imread('Crack.jpg');
3
4 % Downsize the image to facilitate faster computation
  time
5 Im = im2double(imresize(Image,0.2));
6 G = rgb2gray(Im);% Grayscale image
7
8 edges = edge(G,'sobel',0.3);% Find edges using the Sobel
9 filter with a threshold of 0.3
```

Step 3: A window of size ω by ω is centered on each white pixel in the binary edge image. For white pixels near the periphery of the image, it will not be possible use a window without some portion of the window extending beyond the image limits. If this issue is not accounted for, an error will inevitably be encountered during computation. This problem can be eliminated by padding the image by w pixels. Padding involves adding rows and columns of pixels around the border of the image so that even when the window is centered on the most extreme point in the original image (e.g., at the top-left most corner) the additional padding will ensure that the window will not extend beyond the limits of the padded image, and as a result, an error message will be avoided. Having padded the image, the coordinates of the edge points can be found. The padded original, grayscale, and edge images are shown in Figure 5.2.

```
1 w = 101;% w is the window size
2 w_half = round(w/2);% Ensure that this is an integer value
3 % Pad color and grayscale images. Use 'symmetric' option to
4 pads the image with a mirror reflection of border pixels
5 Im = padarray(Im,[w_half, w_half],'symmetric');
6 G = padarray(G,[w_half, w_half],'symmetric');
7 % Pad the binary edge image. Padded pixels are zero by
default
8 edges = padarray(edges,[w_half, w_half]);
9
10 % Find coordinates of all white pixels in the edge image
11 [x y] = find(edges == 1);
```

The pixel at the center of the window is the seed point of the percolated region, R_p. The average gray value of this region is denoted as $\bar{G}_{R_p}$.

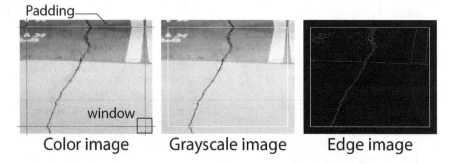

Figure 5.2 **The padded color, grayscale, and edge images of the cracked surface.**

Neighboring pixels are added to the percolated region if their gray level is equal to, or below, the Threshold, T, which is defined as:

$$T = \max \left(\overline{G}_{R_p}, \quad G_{R_n, \min} \right) \tag{5.1}$$

where R_n denotes pixels neighboring the percolated region R_p, and $G_{R_n, \min}$ indicates the minimum gray level (or the darkest pixel) out of all of the neighboring pixels. Essentially, T assumes the maximum value between either the mean intensity value of pixels already in the percolated region, or the minimum intensity value of the pixels neighboring the percolated region. This ensures that there will always be at least one pixel added to the region following each iteration. This process is carried out for w iterations. It can be performed in MATLAB® as follows:

```
1 [M N e] = size(Im);% Find image size
2 output = zeros(M,N);% Initialize output image
3 se = strel('square',3);% Structural element for growing
  region
4
5 % Iterate over all white points in binary edge image
6 for i = 1:1:length(x)
7
8 ptx = x(i); pty = y(i);
9 % Window is centered on (ptx,pty)
10 Window = G(ptx-w_half:ptx+w_half,pty-w_half:pty+w_half);
11 WindowRGB = Im(ptx-w_half:ptx+w_half,pty-w_half:pty+w_half,:);
12
13 %Initialize seed region as w x w array of 0s. Center
   pixel = 1
14 SeedReg = zeros(w_half*2+1);% Initially, s
15 SeedReg (w_half+1,w_half+1) = 1;% The seed point
16
17 % Begin percolation
18 for s = 1:w_half
19 % Candidate pixels to add to the seed region
20 New_Pixels = imdilate(SeedReg,se)-SeedReg;
21
22 % Set threshold as either the mean of current pixels in
   seed
23 region, or the min of candidate pixels, whichever is
   greater
24 T= max([mean(Window(SeedReg~=0)),min(Window(New_Pixels==1))]);
25 % New pixels are added to region if they are <= the
   threshold
26 New_Pixels = (Window<=T).*New_Pixels;
27 % The seed region is expanded to include new pixels
28 SeedReg = max(New_Pixels,SeedReg);
```

```
29 end
30
31 % Classify regions as representing a crack or
   non-crack.
32 IsCrack = classifyCrack(SeedReg, WindowRGB);
33 if IsCrack == true% If it is a crack, add to the output
   image
34 output(ptx-w_half:ptx+w_half,pty-w_half:pty+w_half) = ...
35 output(ptx-35 w_half:ptx+w_half,pty-w_half:pty+w_half)+SeedReg;
36 end
37 end
38
39 % Crop outer parts of image to undo effect of padding
40 output = output(w_half+1:end-w_half,w_half+1:end-w_half);
41 Im = Im(w_half+1:end-w_half,w_half+1:end-w_half,:);
42 %Display final result
43 figure, imshow(output,[])
```

If the seed point is located just outside the crack, the percolated region should gravitate toward the interior of the crack and continue to proceed along the crack line. Examples of how a seed point grows for a sub-image featuring a crack and a sub-image without a crack are presented in Figure 5.3. For the crack region, the seed point grows along the interior of the crack. After the final iteration, the percolated region has a long narrow structure that coincides with the dark crack. For the non-crack region, the seed point percolates outward in all directions and the final shape has no well-defined structure.

Step 4: The process outlined in step 3 is repeated until the percolated region, R_p, reaches the boundary of the window. Percolation ceases and the resulting percolated shape is fed into the classification stage.

Step 5: The first stage of the classification phase involves evaluating the shape of the percolated region. This is done by calculating the eccentricity parameter Ecc. The eccentricity is computed based on the eccentricity of the ellipse that has the same second-moments as the percolated region. The eccentricity is the ratio of the distance between the foci of the ellipse and its major axis length. It has a value of between 0 and 1. An ellipse with an eccentricity of 0 is a circle, while an ellipse whose eccentricity is 1 is a line segment. A long, narrow crack would be expected to have an eccentricity that is closer to one since cracks resemble line segments. With this in mind, percolated regions having an eccentricity greater than some threshold (greater than 0.9 in our case) are retained and are elected to undergo one more classification stage.

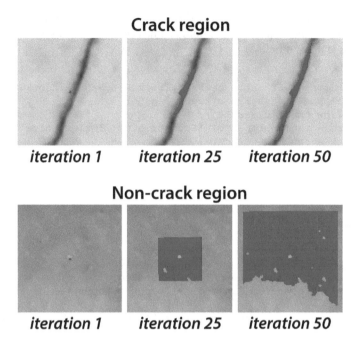

Figure 5.3 Examples of how a seed point grows for a sub-image featuring a crack and a sub-image without a crack.

The second part of the classification phase involves checking the pixel intensity values on either side of the percolated region in order to discount cases where the potential crack has an increased likelihood of being a false alarm, perhaps arising from some innocuous line boundary in the scene. This is achieved by dilating the percolated region in a perpendicular direction on both sides by a distance, w, and comparing the mean of the intensity values from the original image, Im, on both sides, S_1 and S_2, respectively. If the difference between the mean of the pixel values from each side, denoted as C, is greater than a preselected threshold, then the percolated region is rejected. C is expressed as:

$$C = \left| \frac{1}{n_{S_1}} \sum Im(s) - \frac{1}{n_{S_2}} \sum Im(t) \right| \quad s \in S_1,\ t \in S_2 \tag{5.2}$$

where n_{S_1} is the total number of pixels on one side of the crack while n_{S_2} is the number of pixels on the other side.

Step 6: Retained percolated regions from the classification stage are added to the binary output image. The code for the classification process is shown below:

```
1 function [IsCrack] = classifyCrack(SeedReg, WindowRGB)
2 % Compute statistics of seed region
3 cc = bwconncomp(SeedReg);
4 stats = regionprops(cc, 'Eccentricity','Orientation');
5 Ecc = stats.Eccentricity;
6 ang = stats.Orientation;
7
8 IsCrack = false;% Initialize output tag
9
10 if Ecc > 0.90%A crack-like region will have high
eccentricity
11
12 SE = strel('line', size(SeedReg,1), ang+90);
13 % Extract regions on either side of crack
14 Sides = imdilate(SeedReg,SE)-SeedReg;
15 cc2 = bwconncomp(Sides,4);% Compute properties of the
regions
16 stats = regionprops(cc2, 'PixelList');
17
18 % Characterize the color composition of each region
19 MeanRGBSide1 = mean(WindowRGB(stats(1).PixelList,1:3));
20 MeanRGBSide2 = mean(WindowRGB(stats(2).PixelList,1:3));
21 %See how similar the color of both regions are to one
another
22 C=mean(abs(MeanRGBSide1 - MeanRGBSide2));
23
24 % If C < 0.1, we consider the region to represent a crack.
25 % Otherwise, there is a high chance the region is a
   non-crack 26 boundary between two distinct surfaces/
   objects
27 if  C < 0.1
28 IsCrack = true;
29 end
30 end
31 end
```

Examples of sub-image classification are presented in Figure 5.4. In order for a sub-image to be classified as a crack, it must have a high eccentricity (i.e., a narrow structure) and both sides of the crack must be similar to one another (i.e., the C value must be low). This scenario is depicted for the first sub-image in Figure 5.4; however, the second and third sub-images are rejected for having a low eccentricity and a high C value, respectively.

All of the sub-images that are considered to represent a section of the crack are added to an output image. Figure 5.5 shows this output image overlaid on the original image.

Eccentricity = 0.97 ✓   Eccentricity = 0.64 ⊗   Eccentricity = 0.99 ✓
C = 0.01 ✓   C = 0.02 ✓   C = 0.12 ⊗

Crack→true   Crack→false   Crack→false

Figure 5.4 Sub-image classification scenarios.

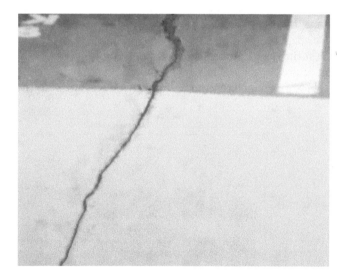

Figure 5.5 Detected crack (in red) overlaid on the original input image.

5.3 PERFORMANCE EVALUATION UNDER VARIOUS CONDITIONS

This section presents a collection of images of cracks that were photographed in both underwater and above water settings. The images present numerous challenges such as lighting complexities, poor visibility, awkward view angles, etc., which will test the limits of the crack detection technique described in the previous section. Following application of the crack detection technique on these images, the results will be presented, and a discussion will be carried out around the influence of various factors on the performance of the crack detection technique.

5.3.1 Test imagery

The test imagery includes both underwater and above water imagery. The images captured in the underwater setting feature a real cracked concrete specimen for varying controlled levels of turbidity and lighting (Figure 5.6). The quality of these photographs is assumed to be chiefly affected by luminosity, sharpness (focus accuracy), contrast, and noise. These quality factors are directly related to the on-site operating conditions, of which lighting and turbidity are the most influential.

Turbidity is defined as the cloudiness in a liquid caused by the presence of suspended solids (Kirk, 1985). These suspended solids scatter and absorb light and therefore reduce visibility. The level of turbidity may be exacerbated by interference during the inspection process, such as from a mechanical source like a boat or human contact with the river bed, which may aggravate and disturb the sediment.

Lighting also plays a pivotal role in achieving good visibility. Ambient light may be sufficient for near-surface inspections; however, it is unlikely to be sufficient at greater depths at which point artificial light sources become necessary. These artificial light sources may introduce luminous complexities

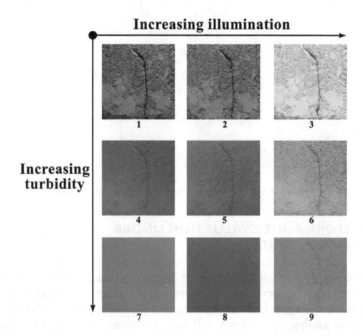

Figure 5.6 Cracked concrete specimen under varying lighting and turbidity conditions. Columns: Low, Medium, High Light. Rows: Low, Medium, High Turbidity.

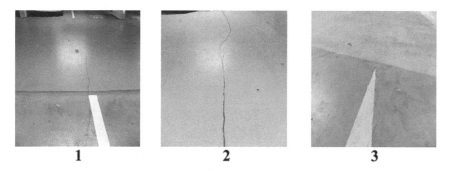

Figure 5.7 Cracks in a conventional above-water setting.

such as "bright-spots," whereby areas of high light intensity may be focused at a spot that may fool the crack detection algorithm. In this section, three levels for turbidity were chosen (0 NTU [clear water], 6 NTU, and 12 NTU) and three lighting levels were used (100 Lux, 1000 Lux, and 10000 Lux).

Although underwater scenes generally provide some of the poorest visibility levels, interpreting images from above-water scenes can also be severely affected by some adverse conditions. It is thus important that the influence of some of the more familiar challenges for crack detection should be investigated. Figure 5.7 shows three images of cracks, each presenting one or more challenges, namely, (1) road markings, (2) illumination complexities (i.e., bright-spots), and (3) monochromatic scene with vague cracks.

5.3.2 Results

The performance of the percolation-based crack detection technique will be evaluated using the receiver operating characteristic (ROC) space. The ROC space provides a common and convenient tool for graphically characterizing the performance of NDT techniques, and its usage has been extended to image detection (Pakrashi et al., 2010). ROC analysis is described more thoroughly in Section 8.2 of this book. For now, it is sufficient to say that the performance points are plotted in the receiver operating characteristic (ROC) space by treating the detection rate (DR) as the vertical coordinate and the misclassification rate (MCR) as the horizontal coordinates.

The detection rate is the proportion of pixels representing a crack that are correctly detected, while the misclassification rate is the proportion of all incorrectly classified pixels to the total number of pixels in the entire image. The detection rate and the misclassification rate are determined by comparing the detected cracks with a visually segmented image. The visually segmented image is created by a human observer who must manually identify the cracks in an image. This visually segmented image acts as the control as it is assumed it shows the true extent of cracks in the image. The visually segmented image only needs to be created when it is wished to gauge the performance levels of the technique under scrutiny. The crack detection technique is applied to the photographs presented in Figure 5.6 and the results of the application are shown in Figure 5.8. The results for the above-water images are shown in Figure 5.9.

The performance of the crack detection technique is summarized in Table 5.1, and the associated performance points are plotted in the ROC space in Figure 5.10.

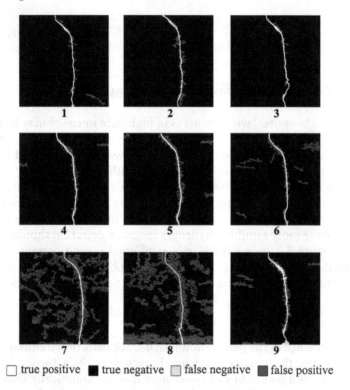

☐ true positive ■ true negative ☐ false negative ■ false positive

Figure 5.8 Detected cracks corresponding to images (1–9) in Figure 5.6.

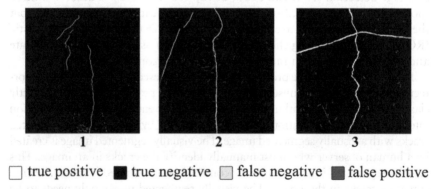

☐ true positive ■ true negative ☐ false negative ■ false positive

Figure 5.9 Detected cracks for the above-water images (1–3) in Figure 5.7.

Table 5.1 Performance of the crack detection technique

Image	Condition	(DR)	(MCR)	δ
Underwater cracks				
I	Low Light, Low Turbidity	89.8%	1.7%	0.10
2	Medium Light, Low Turbidity	87.4%	2.5%	0.13
3	High Light, Low Turbidity	62.6%	1.5%	0.37
4	Low Light, Medium Turbidity	76.3%	2.5%	0.24
5	Medium Light, Medium Turbidity	86.6%	3.7%	0.14
6	High Light, Medium Turbidity	68.6%	4.2%	0.32
7	Low Light, High Turbidity	80.0%	21.3%	0.29
8	Medium Light, High Turbidity	94.0%	26.8%	0.27
9	High Light, High Turbidity	78.1%	4.2%	0.22
Above-water cracks				
I	Road markings	73.0%	0.4%	0.27
2	Illumination complexities	69.2%	0.9%	0.31
3	Monochromatic scene	78.1%	0.6%	0.22

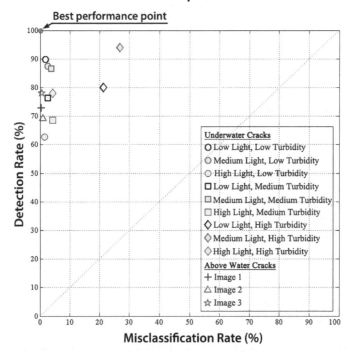

Figure 5.10 Evaluation of crack detection technique through the use of performance points in the ROC space.

It may be observed from the detected regions in Figure 5.8 that the ability of the detection technique is strongly related to the degree of clarity in the original images (Figure 5.6). Unsurprisingly, images that feature poor visibility conditions produced poor performance levels, while on the other hand, images featuring clear and sufficiently lit scenes produced quite good detection results.

At the same time, however, the results suggest that an overly bright light source can be harmful to the detection accuracy. The nature of underwater lighting means that an overly bright light source will induce a bright-spot at the center of the light beam, which will gradually fall-off, resulting in a non-uniform scene. It is this non-uniformity in lighting that the camera cannot effectively compensate for. An example of this can be seen in Figure 5.6(3) where the strong lighting creates a bright area on the surface of the specimen. This washes out some of the detail, thereby preventing a portion of the crack from being detected.

A clear trend that emerges from an analysis of Table 5.1 is that the misclassification rate rises with increasing turbidity. However, the relationship does not appear to be linear. Instead, there is a gradual increase in MCR values from the low to the medium turbidity levels, while there is a pronounced increase between the medium and high turbidity levels. This suggests that as the turbidity approaches the operating limits there is a rapid deterioration in performance. This is especially evident for the high turbidity images in Figure 5.6 (7 and 8) where there is a high degree of false positives contributing to high MCR values in comparison to the lower turbidity levels.

An exception to the decline in performance associated with increasing turbidity and excessive lighting is in Figure 5.6(9). In this case, the high turbidity partially mitigates the high lighting through absorption and diffusion, which limits the formation of a bright spot. Overall, the best results are obtained for the low and medium light levels at the lowest turbidity, and the medium lighting level at the medium turbidity level (Figure 5.6, images 1, 2, and 5).

Analysis of the results in Table 5.1 and the corresponding performance points in the ROC space (Figure 5.10) reveals that each image performs quite consistently in spite of the varying challenges faced in each image. The first image (Figure 5.7(1)) demonstrates the usefulness of employing the classification criteria, which rejects "cracks" having dissimilar sides. It may be observed from the image of the detected cracks (Figure 5.9(1)) that there are very few false positives, even though the boundary between some road markings bears a resemblance to a crack (fine-structured and darker than the surroundings).

The bright spot in Figure 5.7(2) has a deleterious effect on the detection accuracy as the crack remains partially undetected coinciding with this spot. This suggests that attempts to minimize the glare through careful choice of camera viewing angle/position during the on-site image acquisition stage would be worthwhile.

The cracks in Figure 5.7(3) appear quite vague at certain places. There are also numerous instances of staining on the floor that may mislead the crack detection algorithm. As such, it may be expected that the DR would suffer and there would be a high MCR; however, the DR remains reasonably high and the MCR remains low. This may be attributed to percolation-based methods that have been recognized as being well-suited for detecting ambiguous cracks (Wang & Huang, 2010).

5.4 EXTRACTING PHYSICAL PROPERTIES OF DETECTED CRACKS

One of the most straightforward ways of estimating the physical size of detected cracks is by placing an object of known dimensions somewhere close to the crack. This allows a scale factor to be established that relates pixel units to real-world units such as millimeters (refer to Section 3.4 for carrying out this process). In the case of cracks, there are some additional precautions and factors that should be considered. First, it is important that the photographer tries to avoid casting shadows on the crack, which can impair crack detection. For above-water situations where a tripod can be employed, the tripod should be set up in such a way that the shadows do not overlap with the cracks insofar as reasonably possible. On the same note, the scale object should be placed somewhere close to the crack without obscuring or casting shadows on the crack.

Second, it is important to bear in mind that the width of cracks is typically in the order of just a few pixels. If the cracks under inspection are particularly narrow (e.g., hairline cracks) then either a high-resolution camera should be used or a close-up view of the crack should be captured so that fine details can be resolved and the crack is discernible in the acquired imagery.

Third, the camera should be orientated perpendicular to the cracked surface. This reduces the problem of perspective differences where sections of a crack that are close to the camera appear wider than sections of the crack that are farther away from the camera. These factors are illustrated in Figure 5.11.

5.5 SUMMARY

Adopting an image-processing based approach to automatically count and quantify the length and width of cracks can enhance inspections and, in turn, lead to monetary savings or more frequent inspection cycles. This chapter describes an image-processing approach to efficiently and objectively detect cracks. The approach employs a percolation-based method that is applied to locations in the image where there is a large color change,

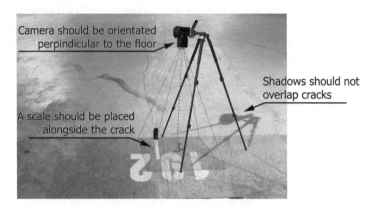

Figure 5.11 **Ideal camera set-up for extracting physical properties of cracks.**

or pixel intensity gradient, suggestive of edges or surface irregularities like cracks. A classification criterion is introduced that considers the pixel intensity values on either side of a crack. Detected regions that match the expected shape of a crack yet have disparate sides are rejected based on the assumption that they represent non-harmful edge boundaries.

Understanding how the operating conditions affect the performance of a crack detection technique is of high practical importance. The performance of the technique is investigated under a range of underwater visibility conditions. The method is also applied to above-water scenes, where a number of other challenges are encountered, such as bright spots, difficult viewing angles, road markings/tire marks, and low-contrast monochromatic scenes.

Finally, some practical guidance on how to capture photographs of cracks is provided with a view on how to extract physical properties of the crack. Overall, this chapter facilitates inspectors when deciding on the appropriateness and implementation of an image-processing based crack detection technique, under a given set of environmental circumstances.

REFERENCES

Cha, Young-Jin, Wooram Choi, and Oral Büyüköztürk. "Deep learning-based crack damage detection using convolutional neural networks." *Computer-Aided Civil and Infrastructure Engineering* 32, no. 5 (2017): 361–378.

Choudhary, Gajanan K., and Sayan Dey. "Crack detection in concrete surfaces using image processing, fuzzy logic, and neural networks." In *Advanced Computational Intelligence (ICACI), 2012 IEEE Fifth International Conference on*, pp. 404–411. IEEE, 2012.

Kirk, John T. "Effects of suspensoids (turbidity) on penetration of solar radiation in aquatic ecosystems." In *Perspectives in Southern Hemisphere Limnology*, pp. 195–208. Springer, Dordrecht, 1985.

Khanfar, Aws, Mohammed Abu-Khousa, and Nasser Qaddoumi. "Microwave near-field nondestructive detection and characterization of disbonds in concrete structures using fuzzy logic techniques." *Composite Structures 62*, no. 3–4 (2003): 335–339.

Pakrashi, Vikram, Franck Schoefs, Jean Bernard Memet, and Alan O'Connor. "ROC dependent event isolation method for image processing based assessment of corroded harbour structures." *Structures & Infrastructure Engineering 6*, no. 3 (2010): 365–378.

Sinha, Sunil K., and Paul W. Fieguth. "Automated detection of cracks in buried concrete pipe images." *Automation in Construction 15*, no. 1 (2006): 58–72.

Wang, Pingrang, and Hongwei Huang. "Comparison analysis on present image-based crack detection methods in concrete structures." In *Image and Signal Processing (CISP), 2010 3rd International Congress on*, vol. 5, pp. 2530–2533. IEEE, 2010.

Yamaguchi, Tomoyuki, and Shuji Hashimoto. "Improved percolation-based method for crack detection in concrete surface images." In *Pattern Recognition, 2008. ICPR 2008. 19th International Conference on*, pp. 1–4. IEEE, 2008.

Chapter 6

Surface damage detection

6.1 INTRODUCTION

Sophisticated image-processing techniques are required to reliably detect anomalies on the surface of infrastructural elements, especially in challenging underwater scenes. Adopting such techniques as part of the inspection regime brings about several benefits. While human inspectors are adept at identifying damage, the task of manually measuring, describing, and recording damage on-site is time-consuming and inexact. Using image-based techniques, physical properties of detected damages can be readily extracted with minimal human supervision, and damage descriptions are quantitative rather than qualitative. The quantitative nature of the data is helpful when estimating the rate of propagation of damage; inspectors can photograph damage and return to the site several years later and determine if the damage is getting worse compared to image analysis measurements recorded from past inspections. Moreover, adopting image-based methods means that inspectors can spend less time on-site, which has significant practical, financial, and safety implications. Furthermore, the remote nature of image collection means that inaccessible parts of a structure can be assessed, and normal operation of the structure does not need to be interrupted.

Detection of 2D damage features can be carried out by using color intensity based methods or texture analysis based methods. Naturally, the techniques in each class are suited to different applications, depending largely on whether the damage instances are more separable from the background based on color or on texture. This chapter describes methodologies for carrying out damage detection using both color and texture approaches and provides some insight into what technique to use in various situations and for the various types of damage encountered in the marine environment.

6.2 TYPES OF DAMAGE ENCOUNTERED
IN MARINE ENVIRONMENT

Structures in marine environment are susceptible to aesthetic, functional, or structural degradation. The degradation process is driven by a number of physical and chemical actions. The physical actions include loading from wind, waves and currents, impact damage from floating objects, and cycles of freezing and thawing that cause expansion and cracking over a period of many years. The chemical actions include corrosion, biofouling and microbially influenced corrosion, reactions between sea-water constituents on cement hydration products, crystallization of salts within concrete, and alkali-aggregate expansion if reactive aggregates are present (Liu, 1991). Once damage is initiated by any of these processes, structures tend to become more susceptible to further attack. For instance, damage to concrete structures can reduce the permeability, which exposes the structure to deeper and more incisive attacks by various processes.

The visual types of damage include rust staining and spalling (corrosion damage), leaching and cement matrix breakdown (chemical attack), loss of section (erosion/abrasion damage), cracking and deflections (structural damage), and map cracking and exudations (alkali-aggregate reactivity). These damage forms are shown in Figure 6.1.

Many damage forms found in marine environment are often accompanied by a perceivable change in color (e.g., corrosion, leaching, etc.). In such cases, color-based techniques would be a natural choice for detecting damage. However, there are also numerous damage forms that are similar in color to the non-damaged surface yet differ in terms of their textural composition (e.g., spalling, honeycombing, erosion, etc.). Texture analysis approaches are especially useful for detecting these types of damages. If the best choice of damage detection algorithm is still not clear after some thought, the underwater image repository discussed in Chapter 8 may be consulted, which allows various techniques to be quantitatively evaluated on a variety of damage types and a variety of on-site visibility conditions.

6.3 COLOR BASED DAMAGE
DETECTION TECHNIQUES

Color based segmentation algorithms may be grouped into four major categories: thresholding, edge detection using gradient information, region growing, and hybrid methods (Abdel-Qader et al., 2008). Existing literature contains a variety of these segmentation methods applied in the domain of NDT. Many of these methods are designed for a particular application such as the detection of weld defects (Yazid et al., 2011), and/or for particular image sources such as optical (Yazid et al., 2011), thermal (Huang & Wu, 2010), ultrasonic (Molero et al., 2012) and radiography (Kasban et al., 2011).

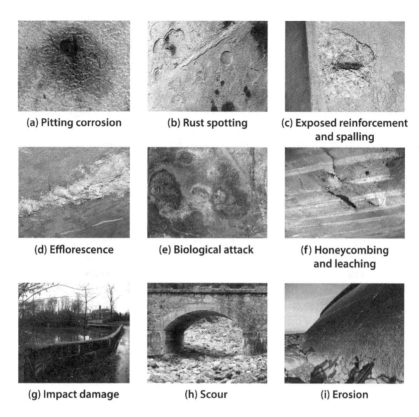

(a) Pitting corrosion

(b) Rust spotting

(c) Exposed reinforcement and spalling

(d) Efflorescence

(e) Biological attack

(f) Honeycombing and leaching

(g) Impact damage

(h) Scour

(i) Erosion

Figure 6.1 Common types of damage encountered in marine environment: (a) pitting corrosion, (b) rust spotting, (c) exposed reinforcement and spalling (credit: Achim Hering), (d) efflorescence (credit: Achim Hering), (e) biological attack, (f) honeycombing and leaching, (g) impact damage, (h) scour, and (i) erosion.

As such, while these techniques may be effective for their designated purposes, they are understandably unlikely to perform well when applied to richly detailed, high-resolution optical images of a broad range of surface types and damage forms in complex natural scenes. There exist very few studies that have developed powerful image processing techniques to cater for the detection of damage in challenging circumstances. Thus, the emphasis lies in the development of a new technique that can characterize features of interest in natural scenes with credibility (Naccari et al., 2005).

6.3.1 Surface damage detection method

This section describes a technique that is designed to detect a broad range of damage forms on the surface of civil infrastructure. For demonstration purposes, the technique is illustrated on a corroding infrastructure component in a harbor facility. The technique is also applied to several other damage forms, where it is shown that it easily extends beyond the application presented and may be considered an effective and versatile general-purpose detection technique. The described technique consists of several stages. To begin with, the technique is shown some training examples of damaged and non-damaged regions. The image of interest is pre-processed through a color reduction scheme in order to form closed geometries corresponding to objects in a scene. Statistical properties are then calculated for each of the closed geometries. These statistical properties serve as input to a clustering based filtering phase where closed geometries having statistical properties characteristic of damaged regions are retained, while closed geometries that have statistical properties characteristic of non-damaged regions are discarded. Extracting statistical properties from a region, rather than processing raw pixel data directly, is often better suited and more robust to the noisy data found in underwater images (Giralt et al., 2013; Li et al., 2013). Following initial region classification, pattern recognition and classification techniques like support vector machines (SVMs) are used to classify pixels at the boundary of damaged regions to improve their size and shape characteristics. Combining each of these constituent phases in an effective manner creates a powerful yet expeditious detection algorithm. The low complexity of the clustering based filtering technique is complemented by the strategic application of the high complexity SVMs. A flowchart illustrating the order of the stages is presented in Figure 6.2.

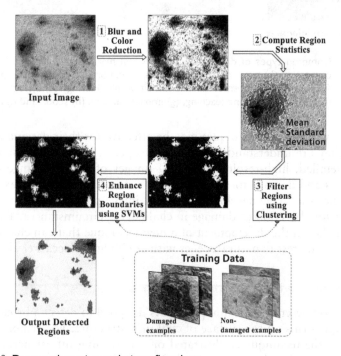

Figure 6.2 **Damage detection technique flowchart.**

The first step is to "show" the algorithm labelled training examples of damaged and non-damaged regions. To avoid underfitting, the training examples should consist of several representative damaged and non-damaged regions, especially in cases where the appearance of the same type of damage can vary markedly. In this demonstration, for the sake of conciseness, only one damaged region and one non-damaged background region will be used as training data. The following MATLAB® code shows how to interactively draw a polygon around the regions of interest and extract the pixel values enclosed in the polygon. The process and output of this stage is illustrated in Figure 6.3.

```
1 % Read in image
2 TrainingImage = imread('TrainImage.tif');
3
4 % Display image and select damaged region
5 imshow(TrainingImage),title('Select Damaged Region');
  hold on
6 Damaged_Region = roipoly(TrainingImage); hold off
7
8 % Select non-damaged region
9 imshow(TrainingImage),title('Select Non-Damaged Area');
  hold on
10 NonDamaged_Region = roipoly(TrainingImage); hold off
11
12 % Display binary images of damaged and non-damaged
   regions
13 figure, imshow(Damaged_Region)
14 figure, imshow(NonDamaged_Region)
15
16 % Extract gray level values in damaged and non-damaged
   regions
17 TrainingGray = rgb2gray(TrainingImage);
18 DamPixels = TrainingGray(Damaged_Region==1);
19 NonDamPixels = TrainingGray(NonDamaged_Region==1);
```

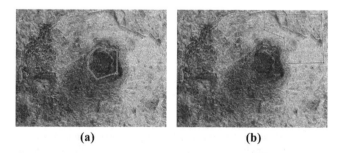

(a)   (b)

Figure 6.3 Selecting training data: (a) polygon around a damaged region, (b) polygon around a non-damaged region.

Once the technique has been shown examples of damaged and non-damaged regions, it can be applied to new images to classify instances of damage.

6.3.1.1 Identification

The first stage of the damage classification process involves performing blurring and color reduction operations on the input image. Blurring serves to attenuate high spatial frequencies and causes neighboring pixels to assume similar pixel intensity values. When this is followed by a color reduction operation, the effect is that contiguous regions emerge that correspond to the same object or region in the image. All pixels in a region share the same quantized pixel intensity value, which makes it straightforward to extract properties from the region and carry out further processing. Blurring is discussed in detail in Section 4.3. In short, blurring is achieved by convolving the grayscale image with a filter. Common blurring filters include an equally weighted filter, whereby the new gray level of a pixel is simply the average of adjacent pixels that are within the field covered by the filter, or, as is the case in this demonstration, a Gaussian filter, whereby neighboring pixels have a greater contribution to the value of the output pixel than pixels that are a farther distance away. The result of blurring is shown in Figure 6.4(b).

The color reduction operation involves histogram equalization followed by color quantization. Histogram equalization (see Section 4.3) serves to transform the intensity values in an input grayscale image in such a way that the gray levels in the output image are approximately evenly distributed across all discrete intensity levels. This step is introduced as a way to condition the data and to help ensure that each quantized color level is roughly equally represented. Color quantization apportions the gray levels into n discrete bins. This parameter may be optimised using a ROC-based optimization framework/trial and error approach. In this demonstration, a value of $n = 5$ is used that creates a quantized image

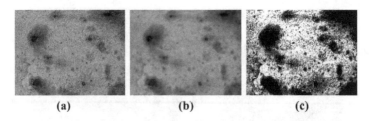

(a)   (b)   (c)

Figure 6.4 (a) Grayscale image, (b) grayscale image after blurring operation, and (c) outcome of this stage—a quantized grayscale image where contiguous regions that share the same gray level value are present.

with 5 discrete gray levels (Figure 6.4(c)). This process is outlined in the following code:

```
1  % Read in image and convert it to grayscale
2  Im = im2double(imread('Corrosion.jpg'));
3  Gray = rgb2gray(Im);
4  % Blur to attenuate high spatial frequencies
5  BlurFilter = fspecial('gaussian',[20,20],10);
6  BlurIm = imfilter(Gray,BlurFilter, 'replicate');
7
8  % Perform histogram equalization
9  HisteqIm = histeq(BlurIm);
10
11  % Quantise image to 5 discrete levels
12  n = 5;% NumberOfThresholds
13  thresh = multithresh(HisteqIm,n-1);
14  seg_I = imquantize(HisteqIm,thresh);
```

It may be observed in Figure 6.4(c) that many regions formed from this stage are of a negligible size. These regions either tend to represent spurious regions or represent damaged regions that are smaller than the minimum defect size and, thus, are considered to present an insignificant degree of damage. For computational parsimony and classification accuracy purposes, closed geometries below a certain size are not considered for further analysis. This is a well-established decision that is taken in any damage detection process. For ease of application, it may be convenient for the user to represent the threshold area as a function of the overall image size. For instance, it could be specified that closed geometries less than 1% of the total image area should be discarded. A priori knowledge of the damage type and its relationship to the decision process (repair, detailed inspection, etc.) may be used as a factor in choosing the threshold area. The remaining closed geometries are classified by a clustering based filtering approach.

6.3.1.2 Clustering-based filtering

When capturing imagery on-site in harsh conditions, the uncertainties of measurement reach a high level. As such, classifying individual pixels (or whole regions) based on raw color data is unreliable, and it becomes necessary to resort to statistical estimates. This step computes the mean and standard deviation of pixel values in each of the regions obtained in the previous step. The region properties are found by iterating over all discrete intensity levels of the quantized grayscale image, identifying contiguous regions at each quantized level, and then consulting the grayscale version of the original input image to obtain the gray pixel distribution. From this, the mean and standard deviation values are computed.

The centroid of each region is given by the point $(\mu,\sigma)_o$ where μ is the mean and σ is the standard deviation of the grayscale pixel distribution within the oth closed geometry. The mean of each region is computed as per Equation 6.1:

$$\mu_o = \frac{1}{n_{R_o}}\left(\sum_{t=1}^{n_{R_o}} v_{t,o}\right) \qquad t \in R_o \tag{6.1}$$

while the standard deviation is given by Equation 6.2:

$$\sigma_o = \sqrt{\frac{1}{n_{R_o}}\sum_{t=1}^{n_{R_o}}(v_{t,o}-\mu_o)^2} \qquad t \in R_o \tag{6.2}$$

where $v_{t,o}$ denotes the intensity value for the pixel with index t within the oth region, while n_{R_o} denotes the total number of pixels in the oth region. Standard deviation provides a measure of the variation or dispersion of a set of data values. It is helpful for assessing the homogeneity in a region. While standard deviation gives a measure of spread, it fails to provide information about the center of a distribution. Using standard deviation in conjunction with the mean is a more effective way to describe the pixel distribution within each closed geometry; representing a closed geometry solely based on the mean is susceptible to error as closed geometries with disparate pixel distributions may yield similar mean values. Introducing standard deviation offsets this issue and creates a more well-rounded description of each closed geometry.

Given a set of closed geometries $(R_o = R_1, R_2,..., R_J)$, the clustering algorithm aims to partition the J observations into two sets $S = \{S_1, S_2\}$ such that the Euclidean distance between the centroid of R_o and the centroid of the set that it is assigned to is minimized. S_1 corresponds to the cluster representing damaged regions while S_2 represents the undamaged cluster. The centroid of the oth closed geometry, R_o, is given by $(\mu, \sigma)_o$. The centroids of the damaged and undamaged clusters are obtained from the training data (manually selected regions displayed in Figure 6.3).

R_o is assigned to the set that minimizes the Euclidean distance between the observation centroid, that is, $(\mu, \sigma)_o$, and cluster centroid as per Equation 6.3:

$$
R_o \in \begin{cases} S_1, & \text{if } \sum_\kappa \left\| R_{o,\kappa} - R_{damaged,\kappa} \right\|^2 \leq \sum_\kappa \left\| R_{o,\kappa} - R_{undamaged,\kappa} \right\|^2 \\[2ex] S_2, & \text{otherwise} \end{cases} \tag{6.3}
$$

where κ denotes the index of the elements in R_o. This is expressed in the following code:

```
1 % Iterate over all quantized gray levels and extract
  properties
2 s = [];
3 for i = 1:n% n is the number quantized gray levels
4 CC = bwconncomp(seg_I==i);
5 s  = [s;regionprops(CC,'area','PixelIdxList')];
6 end
7 areas = cat(1, s.Area);% Store region areas in an array
8 minArea = 1000;% Minimum defect size threshold
9
10 % Compute mean and standard deviation of training data
11 % Stats of damaged training region
12         DamagedStats(1,1) = mean(DamPixels);
13         DamagedStats(1,2) = std(DamPixels);
14 % Stats of non-damaged training region
15         NonDamagedStats(1,1) = mean(NonDamPixels);
16         NonDamagedStats(1,2) = std(NonDamPixels);
17
18 RegionStats = zeros(length(areas),2);
19 DamagedRegions = zeros(size(Gray));
20 for i = 1:length(areas)
21     if s(i).Area > minArea% Ignore small areas
22         RegionStats(i,1) = mean(Gray(s(i).PixelIdxList));
23         RegionStats(i,2) = std(Gray(s(i).PixelIdxList));
24     if sum((RegionStats(i,:)-DamagedStats(1,:)).^2) <...
25     sum((RegionStats(i,:)-NonDamagedStats(1,:)).^2)
26         DamagedRegions(s(i).PixelIdxList) = 1;
27     end
28   end
29 end
```

Once the closed geometries have been grouped into their respective clusters, there is scope to enhance their size and shape characteristics.

6.3.1.3 Support vector machine enhancement

Following the region based clustering stage, there still exists many damaged pixels around the periphery of the region that remain undetected. Performing SVM in the neighborhood of these regions and then locally supplementing the closed geometries with the SVM pixels produces better-defined features of interest. This is conveyed by comparing a detected region before and after the local application of SVM classified pixels (Figure 6.5(a) and 6.5(b), respectively).

SVMs are used to classify pixels as being either damaged or undamaged based on the intensity values for each color channel. SVM is a supervised learning classifier based on statistical learning theory. The linear SVM is used for linearly separable data using an $(n\text{-}1)$ dimensional hyperplane in n-dimensional feature space (Cortes & Vapnik, 1995; Cristianini & Shawe-Taylor, 2000). This hyperplane is called a maximum-margin hyperplane that ensures maximized distance from the hyperplane to the nearest data points on either side in a transformed space. The linear kernel function is the dot product between the data points and the normal vector to the hyperplane. The kernel function concept is used to simplify the identification of the hyperplane by transforming the feature space into a high dimensional space. The hyperplane found in the high dimensional feature space corresponds to a decision boundary in the input space. The classifier hyperplane is generated based on the previously selected training datasets.

The enhancement process first examines pixels that are immediately adjacent to each retained region, R_j. A pixel is considered to be adjacent to a region if it shares an edge or corner with any pixel on the periphery of that region. SVM classification is applied to these adjacent pixels utilizing their original intensity values $(a_{x,y})$ to classify each of these pixels as representing a damaged surface or not. Pixels that are classified using SVMs as representing damage become a member of the region, R_j. Pixels in the immediate

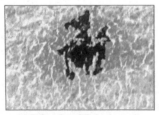 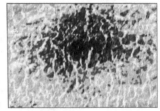

(a) Before Local Enhancement   (b) After SVM Enhancement

Figure 6.5 Close-up of a region, (a) before local enhancement, and (b) detected pixels following SVM classification.

vicinity of the newly identified member pixels of R_j are further subjected to classification using SVMs. This process is repeated until there are no more adjacent damaged pixels that can be added to a region. Alternatively, for greater computational parsimony, the region enhancement phase can be terminated after a set number of iterations. The code for carrying out the SVM boundary enhancement is as follows:

```
1 % Use same number of samples for damaged and undamaged
  data
2 minObs = min(length(DamPixels),length(NonDamPixels));
3 X = [DamPixels(1:minObs); NonDamPixels(1:minObs)];
4 Y = [ones(minObs,1); zeros(minObs,1)];% labels → 1's =
  damage
5 SVMModel = fitcsvm(X,Y);% Train SVM
6
7 se = strel('disk',1);% Structural element for dilating
  regions
8
9 for iter = 1:100% Terminate after 100 iterations
10
11 % Find pixels bordering a region
12 CPixels = imdilate(DamagedRegions,se) -
13 imerode(DamagedRegions,se);
14
15 %Classify these pixels
16 label = predict(SVMModel,Gray(CPixels == 1));
17 % Add newly classified damaged pixels to existing
  regions
18 DamagedRegions(CPixels == 1) = label;
19 end
20 figure,imshow(DamagedRegions)% Display result
```

6.3.2 Technique evaluation and comparison with other methods

For ease of demonstration, the technique has been described using only grayscale information; however, the same methodology can be easily extended to three color channels. This section presents the results obtained by this technique along with several established segmentation techniques. These techniques are Otsu's Method (Otsu, 1979), Chan-Vese Method (Chan & Vese, 2001), Delaunay Triangulation (Cheddad et al., 2008), Region Growing (Adams & Bischof, 1994), and Graph-Based Segmentation (Felzenszwalb & Huttenlocher, 2004). The comparison serves to highlight the qualities and drawbacks offered by each of the various segmentation methodologies. The regions detected using these techniques are shown in Figure 6.6, and their respective performances are quantified in Table 6.1.

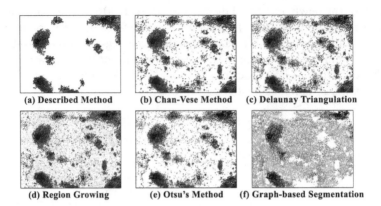

|(a) Described Method|(b) Chan-Vese Method|(c) Delaunay Triangulation|
|(d) Region Growing|(e) Otsu's Method|(f) Graph-based Segmentation|

Figure 6.6 Detected regions from (a) the method described in this chapter, (b) Chan-Vese Method, (c) Delaunay Triangulation, (d) Region Growing, (e) Otsu's Method, (f) Graph-Based Segmentation.

Table 6.1 Comparison of techniques

Segmentation technique	SDR image		
	DR	MCR	δ
Our Method	84%	8%	0.18
Chen-Vese	93%	18%	0.19
Delaunay Triangulation	86%	14%	0.20
Region Growing	89%	17%	0.21
Otsu's Method	86%	14%	0.20
Graph Cutting	78%	39%	0.45

It may be observed from Figure 6.6(a) that the detected regions from the method described in this chapter produce a "cleaner" image that is not contaminated by speckles of spurious regions, which is a trait of the other techniques. Having a "cleaner" image is helpful for many post-processing applications such as labelling/numbering damaged regions. The presence of many small and insignificant spurious regions would hamper this labelling/numbering process and would likely require additional rulesets to be introduced to account for these extraneous pixels, for example, discarding all regions smaller than a specified minimum size threshold.

The Chan-Vese method (Figure 6.6(b)) and Delaunay Triangulation (Figure 6.6(c)) performed quite well, while Otsu's Method (Figure 6.6(e)) also performed reasonably well despite its simple and non-contextual nature. Non-contextual techniques (e.g., thresholding) do not take into account any spatial relationships between pixels in an image, but rather segment pixels at a global level on the basis of some attribute, for example, color intensity. Contextual techniques, on the other hand, do consider spatial relationships. Contextual techniques are often better suited for damage detection as pixels corresponding to a damaged region in an image are spatially connected. The Graph Cutting technique (Figure 6.6(f)) produced poor results, as signified by the larger δ value in Table 6.1 (smaller δ values signify better performance).

6.4 TEXTURE ANALYSIS

Texture is an innate property of surfaces (Haralick & Shanmugam, 1973). There are numerous image processing based techniques that attempt to characterize texture: wavelet analysis (Lu et al., 1997), Laws' texture energy (Choi et al., 2011), First Order Statistics (FOS) (Gill, 1999), GLCM (Gadelmawla, 2004), or hybrid methods (O'Byrne et al., 2013). There are a wide range of damage forms on the surface of infrastructural elements that are similar in color to the non-damaged surface yet are still easily identifiable by human observers as they are markedly different in terms of their textural composition. Such examples include spalling, honeycombing, erosion, etc. Detecting these damage forms using computational methods presents new challenges. Conventional image-based methods that rely on color information will not produce good detection results. Instead, texture analysis based methods provide a more appropriate solution. This section describes and evaluates a texture segmentation approach to detect and classify surface damage on infrastructure elements. The texture analysis approach involves generating feature vectors at each point in the image, which are populated with statistics derived directly from the image as well as from a gray level co-occurrence matrix (GLCM). Support vector machines (SVMs) are used to classify these feature vectors. The technique is successfully applied to a range of images that feature a variety of damage forms.

6.4.1 Methodology

The texture analysis based damage detection algorithm that has been proposed in this chapter involves two main stages. The first stage computes a texture characteristics map from an input color image, and the second stage subsequently classifies pixels as being either damaged or undamaged using SVMs. The overall methodology for the proposed technique is illustrated in the flowchart in Figure 6.7.

6.4.1.1 Texture characteristics map

A texture characteristics feature vector $\{v_f\}_{a,b,c}$ has to be generated for each pixel within the original image, I, for each color channel, c, where f

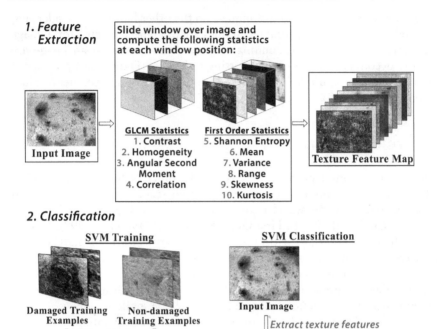

Figure 6.7 **Methodology flow chart.**

indicates the index of the vector element and (a,b) indicates the spatial coordinates of the pixel. The first four elements of $\{v_f\}_{a,b,c}$ are obtained by computing statistics derived from a GLCM. These statistics are (1) angular second moment (ASM), (2) homogeneity, (3) contrast, and (4) correlation. The GLCM is primarily calculated for gray images yet may be readily extended to individual color channels. The remaining six first-order texture features are based on measures calculated from the original pixel values mapped over a range of [0,255]. This range corresponds to 8-bit grayscale images, which are capable of representing 256 (2^8) different intensity levels (or shades of gray). A value of 0 represents black, while a value of 255 represents white. These features are Shannon entropy, mean, variance, range, skewness, and kurtosis.

The feature vector is generated for each pixel using a sliding window, SW, that moves throughout the image and provides the basis for the GLCM statistics and the distributions used for calculating descriptive statistics and Shannon entropy. The window starts at the top left-hand corner of the image and moves horizontally in steps of one pixel until it reaches the end of a row, at which point it progresses onto the leftmost point in the next row. The center is indicated as (a,b) and the size of the window (N-pixel × N-pixel) is optimized for best performance. A trial-and-error approach is used to determine the optimal size. This optimization step may be worthwhile if large batches of images featuring similar damaged surfaces are being processed; however, it was experimentally found that the classification accuracy of the technique was not overly sensitive to the window size. An increase in the size of SW is accompanied by a marginal increase in the overall computational time. In this chapter, a nominal window size of 10 × 10 square pixels was used.

6.4.1.2 GLCM features

The process in which the GLCM is created is illustrated in Figure 6.8. The GLCM is a matrix of frequency values of paired combinations of pixel intensities as they appear in certain specific spatial arrangements within an image or sub-image. The GLCM for each pixel is generated through a sub-image that is a sliding window, SW, centered on the pixel. Combinations

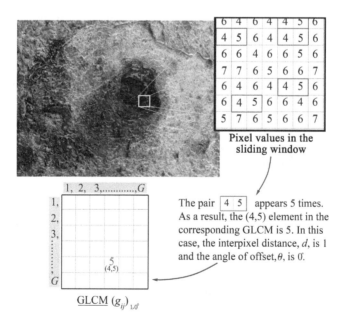

Pixel values in the sliding window

The pair $\boxed{4\ 5}$ appears 5 times. As a result, the (4,5) element in the corresponding GLCM is 5. In this case, the interpixel distance, d, is 1 and the angle of offset, θ, is 0°.

$$\underline{\text{GLCM}}\ (g_{ij})_{1,0°}$$

Figure 6.8 Overview of the GLCM process.

of various pixel pairs within SW were counted and the resulting total was assigned to the g_{ij} in the GLCM, which corresponds to the spatial arrangement of the pixel pairs being summated. The spatial indices i and j of the GLCM match the gray level in the reference pixel and the destination pixel, respectively. The spatial arrangement of the reference pixel and destination pixel in relation to each other in SW are governed by two parameters: the interpixel distance, d, and the angle of offset, θ. The gray levels are defined using integer values between 1 and G. In this chapter, the gray levels are defined on a scale of 1–8 ($G = 8$) instead of a larger scale such as [0,255]. Quantizing in this manner increases computational parsimony at the expense of making the GLCM less sensitive to minute fluctuations in pixel intensity values within the sliding window. Despite this reduced sensitivity, the discriminatory capabilities of the GLCM remain largely unperturbed as perceivable changes in intensity values between neighboring pixels continue to be taken into account, thus creating condensed yet descriptive matrices.

An illustrated example of the creation process for a GLCM is presented in Figure 3.2. In this example, the number of occurrences of pixels with a quantized gray level of 4 and 5 appearing horizontally alongside each other in the sliding window ($d = 1$ and $\theta = 0°$) are computed. The number of occurrences of this pair is then assigned to the (4,5) element in the GLCM corresponding to the chosen value of d and θ. It was experimentally found that paired combinations of intensity values of pixels that are spatially neighboring tend to be more relevant than combinations that involve spatially distant pixels. With this in mind, a value of 1 was chosen for d to ensure a certain level of spatial proximity. The angle along which the interpixel distance is counted is defined as the angles of offset. Four angles for the offset were chosen: $\theta = 0°$, $\theta = 45°$, $\theta = 90°$, $\theta = 135°$. So, this generated a set of 4 GLCMs ($d = 1$; $\theta = 0°$, $45°$, $90°$, $135°$) for each color channel at each pixel location.

The GLCM for each pixel is populated as per Equation 6.4:

$$(g_{ij})_{d,\theta} = \sum_{u=1}^{N} \sum_{z=1}^{N} A \quad \text{where } A = \begin{cases} 1 & \text{if } s_{uz} = i \text{ and } s_{uz}^{d,\theta} = j \\ 0 & \text{otherwise} \end{cases} \tag{6.4}$$

where s_{uz} is the pixel intensity expressed in quantized gray levels for the reference pixel located at row u and column z within the sliding window; $s_{uz}^{d,\theta}$ is the pixel intensity expressed in quantized gray levels for the destination pixel located at an interpixel distance d along an angle θ from the reference pixel. The GLCMs are normalised as per Equation 6.5:

$$p(i,j)_{d,\theta} = \frac{(g_{ij})_{d,\theta}}{N(N-1)} \tag{6.5}$$

Each texture feature determined from the GLCM loosely relates back to some physical textural property of the photographed surface (Baraldi & Parmiggian, 1995). This can be seen in Figures 6.9 and 6.10, which show how the GLCM texture features vary as the textural properties of the illustrated surface also vary. The four texture features, and their physical meaning, are described as follows:

Angular second moment (ASM) represents the uniformity of distribution of gray level in the image and is given by Equation 6.6.

$$(v_{f=1})_{d,\theta} = \sum_{i=1}^{G} \sum_{j=1}^{G} \left(p(i,j)_{d,\theta}\right)^2 \tag{6.6}$$

v_1 ranges from $1/G^2$ to 1. A high value of 1 indicates a uniform image, as may be expected of a smooth painted surface that lacks tonal variation. A low ASM value, on the other hand, may be expected of natural surfaces, which typically have a higher degree of tonal variation.

Homogeneity gives a measure of the similarity of gray levels, as given by Equation 6.7.

$$(v_{f=2})_{d,\theta} = \sum_{i=1}^{G} \sum_{j=1}^{G} m \cdot p(i,j)_{d,\theta} \qquad \text{where } m = |i - j| \tag{6.7}$$

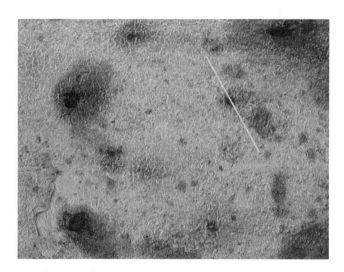

Figure 6.9 Profile line through a corroded region.

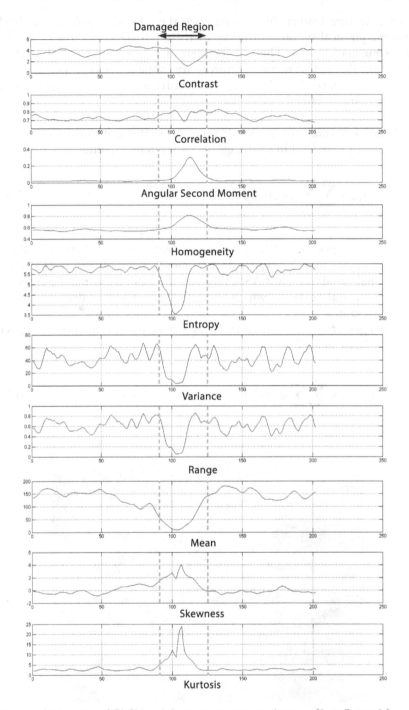

Figure 6.10 Variation of GLCM and descriptive statistics along profile in Figure 6.9.

v_2 ranges from 0 to G-1. A value of 0 indicates a strong similarity of gray levels in the image, as may be expected of a surface that has a constant or periodic form.

Contrast is a measure of the local variations present in an image. The contrast will be high if there is a high amount of variation. Contrast is computed as per Equation 6.8.

$$(v_{f=3})_{d,\theta} = \sum_{m=0}^{G-1} m^2 \left\{ \sum_{i=1}^{G} \sum_{j=1}^{G} p(i,j)_{d,\theta} \right\} \tag{6.8}$$

v_3 ranges from 0 to $(G$-1$)^2$. A value of 0 indicates a uniform image, as may be expected of a smooth manmade material or a painted surface.

Correlation is a measure of the gray level linear-dependencies in an image. Correlation will be high if an image contains a considerable amount of linear structure, which is often the case with manmade objects. Correlation is computed as per Equation 6.9.

$$(v_{f=4})_{d,\theta} = \sum_{i=1}^{G} \sum_{j=1}^{G} \frac{(ij)p(i,j)_{d,\theta} - \bar{\mu}_1\bar{\mu}_2}{\sigma_1\sigma_2} \tag{6.9}$$

where $\bar{\mu}_1$, $\bar{\mu}_2$, σ_1, and σ_2 are the means and standard deviations of the marginal probability matrices, P_1 and P_2, obtained by summing the rows and columns of $p(i,j)_{d,\theta}$, respectively. v_4 ranges from –1 to 1. A value of 1 indicates a perfectly positively correlated image. The denominator in Equation 6.9 will be equal to zero in the case of a perfectly uniform image. This will result in the value of correlation being undefined; however, these undefined values are ignored by the SVM classifier and their influence on the classification accuracy is negligible as these cases tend to appear infrequently, in particular, when working with images of natural scenes.

6.4.1.3 Descriptive statistics and Shannon entropy

The feature vector was further populated by considering five descriptive statistics of the pixel intensity values, along with Shannon entropy. These six features were derived for each pixel using the same sliding window approach employed to calculate the GLCM features. Unlike the GLCM approach, the intensity values used in the distribution adopted the scale [0,255] for several reasons. First, the nature of these first order statistics (FOS) generated directly from the intensity values differed from that of the GLCM statistics as it was the magnitude of the intensity values that was considered and not their frequency of occurrence. As such, it was more important for the intensity values to contain as much information as possible, which required them to be accurately and precisely defined. Having a bigger sample space provided more sensitive information for characterizing texture. Conversely, the GLCM statistics produced more meaningful results by having similar intensity values grouped together as separately counting perceptually close values may understate their prominence in the sliding window. Second, the number of gray-levels employed in the GLCM generation stage directly affected the size of the GLCM, which in turn affected the computational time of the algorithm. The intermediate GLCM generation stage already accounted for a significant portion of the algorithm time, so, for this reason, it was desirable to keep the size of the GLCM to a minimum. Employing quantized intensity values for the descriptive statistics and Shannon entropy, on the other hand, resulted in no perceivable benefits in terms of increased computational efficiency.

As with the GLCM based statistics, each of the FOS in the feature vector describes some aspect of the textural composition in a sliding window. The meaning and contribution of each statistic is discussed. Shannon entropy, v_5, is a statistical measure of the uncertainty associated with a random variable and is given by Equation 6.10.

$$(v_{f=5}) = -\sum_{u=1}^{N}\sum_{z=1}^{N} \tilde{s}_{uz} \log_2 \tilde{s}_{uz} \tag{6.10}$$

v_5 ranges from $-(N^2.\max(\tilde{s}_{uz}).\log_2(\max(\tilde{s}_{uz})))$ to infinite, which for a pixel intensity range of $[0,255]$ becomes $[-2039.N^2,\infty]$.

Mean gives the arithmetic average of the intensity values in a window. It is computed as per Equation 6.11.

$$(v_{f=6}) = \frac{1}{N^2}\sum_{u=1}^{N}\sum_{z=1}^{N} \tilde{s}_{uz} \tag{6.11}$$

v_6 can range from the minimum value of $\tilde{s}_{uz}$, 0, to maximum value of $\tilde{s}_{uz}$, 255.

Variance is a measure of how far a set of numbers is spread out from the mean, as expressed by Equation 6.12.

$$(v_{f=7}) = \frac{1}{N^2}\sum_{u=1}^{N}\sum_{z=1}^{N} (\tilde{s}_{uz} - v_6)^2 \tag{6.12}$$

v_7 ranges from 0 to $\dfrac{\left(\max(\tilde{s}_{uz}) - \min(\tilde{s}_{uz})\right)^2}{4}$, which equates to 1.625×10^4 for the $[0,255]$ range.

Range gives the difference between the maximum and minimum intensity values in the distribution, as represented by Equation 6.13.

$$(v_{f=8}) = \max(\tilde{s}_{uz}) - \min(\tilde{s}_{uz}) \quad \forall(u,z) \tag{6.13}$$

Skewness is a measure of the asymmetry of the data around the sample mean.
An estimate for the skewness is given as per Equation 6.14.

$$(v_{f=9}) = \frac{1}{v_7^{3/2}} \sum_{u=1}^{N} \sum_{z=1}^{N} (\tilde{s}_{uz} - v_6)^3 \tag{6.14}$$

v_9 ranges from $-\infty$ to ∞.

Kurtosis is a measure of the peakedness of a distribution. A positive value for kurtosis indicates that the distribution has a greater peakedness than that predicted by a normal distribution, while a negative value indicates that the distribution is less peaked than predicted by a normal distribution. An estimate for the kurtosis is given by Equation 6.15.

$$(v_{f=10}) = \frac{1}{v_7^2} \sum_{u=1}^{N} \sum_{z=1}^{N} (\tilde{s}_{uz} - v_6)^4 - 3 \tag{6.15}$$

v_{10} ranges from -2 to ∞.

As with the GLCM statistics, undefined values, or infinite values, can result for certain descriptive statistics such as skewness and kurtosis when the intensity values in the window are perfectly uniform, that is, when the standard deviation is equal to zero. The value of entropy may also be undefined in the case of pixel intensities having a value of zero in a given distribution. These undefined values are ignored by the SVM classifier. Their influence on the classification accuracy is negligible, however, as not only do the undefined values tend to appear infrequently, but by having a large feature vector containing a greater number of correctly defined texture measures, their effect is vastly diminished. Moreover, since the sensitivity of each texture measure varies according to the surface type and damage form, having a large feature vector is useful as it ensures that the influence of any texture measure that is ineffective at differentiating between damaged and undamaged regions is offset by other texture features that have a higher sensitivity to regions of contrasting texture. Figure 6.10 plots the GLCM and descriptive statistics along the profile line shown in Figure 6.9, which passes over damaged and undamaged regions.

The code for extracting texture features is presented below:

```
1 % Extract texture related statistics
2 function [FeatureMap] = ExtractFeatures(GrayImage)
3
4 w = 5% window radius
5 [M,N] = size(GrayImage);
6 FeatureMap = zeros(M,N,10);
7 for i = 1+w:M-w
8    for j = 1+w:N-w
9        roi = GrayImage(i-w:i+w, j-w:j+w);
10 glcm = graycomatrix(roi,'Offset',[0 1; -1 1;-1 0;-1
   -1],...
11   'NumLevels',8,'GrayLimits',[]);% Create GLCM
12 stats = graycoprops(glcm,{'Contrast','Correlation',...
13   'Energy','Homogeneity'});% Extract stats from GLCM
14 FeatureMap(i,j,1:4) = [mean(stats.Contrast),...
15 mean(stats.Correlation),mean(stats.Energy),mean(stats.
   Homogeneity)];
16 FeatureMap(i,j,5) = mean(roi(:));
17 FeatureMap(i,j,6) = var(roi(:));
18 FeatureMap(i,j,7) = entropy(roi(:));
19 FeatureMap(i,j,8) = range(roi(:));
20 FeatureMap(i,j,9) = skewness(roi(:));
21 FeatureMap(i,j,10) = kurtosis(roi(:));
22    end
23 end
24
   end
```

The feature map for each texture related statistic for the image in Figure 6.9 is shown in Figure 6.11.

6.4.1.4 Support vector machine classification

Support vector machines are used to classify pixels as being either damaged or undamaged, based on the feature vector assigned to each pixel. SVM is a kernel-based supervised learning classifier, based on statistical learning theory. The goal of SVM is to produce a model that predicts the target values of the test data given only the test data attributes. The first stage in the classification process involves training in the data. Representative samples from damaged and undamaged regions in each image were selected and the texture features corresponding to these regions provided the basis for separation. Each pixel in the image was then classified based on the texture features corresponding to that pixel.

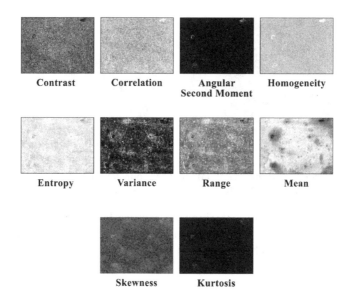

Figure 6.11 Feature map for each of the texture related statistics.

The first step is to train the SVM using supplied labelled training examples of damaged and non-damaged regions. In this demonstration, we will interactively crop out a damaged region and a non-damaged region from an image. The texture related statistics, or texture features, for these two cropped sub-images is then found and the texture features are used to train the SVM. The process and output of this stage is illustrated in Figure 6.12.

```
1 % Read in training image and crop damaged/non-damaged
  regions
2 TrainingImage = im2double(imread('TrainImage2.tif'));
3 TrainingGray = rgb2gray(TrainingImage);
4 DamReg = imcrop(TrainingGray);
5 NonDamReg = imcrop(TrainingGray);
6
7 % Extract texture features for each region
8 [DamVector] = ExtractFeatures(DamReg);
9 [NonDamVector] = ExtractFeatures(NonDamReg);
10
11 % Train SVM
12 DamVector = reshape(DamVector,[],10);
13 NonDamVector = reshape(NonDamVector,[],10);
14 X = [DamVector; NonDamVector];
15 Y = [ones(size(DamVector,1),1); zeros(size
   (NonDamVector,1),1)];
16 SVMModel = fitcsvm(X,Y);
```

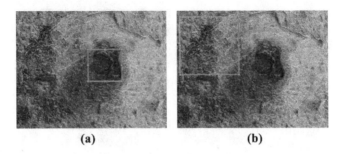

(a)   **(b)**

Figure 6.12 **Selecting training data: (a) cropped out for the damaged region, and (b) cropped out for the non-damaged region.**

Once the SVM has been trained on texture features relating to damaged and non-damaged regions, it can be applied to new images to classify instances of damage. This is achieved as follows:

```
1 % Read in new image and classify
2 Im = im2double(imread('Corrosion.jpg'));
3 Gray = rgb2gray(Im);% Convert to grayscale
4
5 %Extract texture features
6 [FeatureMap] = ExtractFeatures(Gray);
7 [M,N] = size(Gray);
8 output= zeros(M,N);% Instantiate the output binary map
9 w = 5;
10 for i = 1+w:M-w
11     for j = 1+w:N-w
12         FeatureVector(1:10) = FeatureMap(i,j,1:10);
13         %Classify at each pixel location
14         [label, score] = predict(SVMModel,FeatureVector);
15         % label has a value of +1 for damage, 0 for
           non-damage
16         output(i,j) = label;
17         ScoreMap(i,j) = score(1);
18     end
19 end
```

The result of this classification step is an output binary map and a "score" map. The output map shows the points in the image that have been classified as damage (white pixels) and non-damage (black pixels), as shown in Figure 6.13(a). The score map gives a measure of the cost of assigning a pixel to a particular class. The dark pixels in Figure 6.13(b) correspond to points where there is a low cost of assigning these pixels to the damage class.

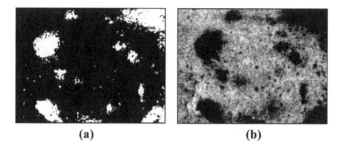

(a)   (b)

Figure 6.13 (a) Detected regions from texture analysis, and (b) a map showing the cost of assigning pixels to the damage class.

6.5 COMPARISON OF COLOR AND TEXTURE BASED METHODS

Naturally, the color and texture based methods that have been described in this chapter are suited to different applications, depending largely on whether the damaged regions are more separable from the background based on color or on texture. This section looks at the performance of the color based method alongside a texture analysis based technique (O'Byrne et al., 2013), in order to give an indication of the performance levels that can be expected when a range of damage forms and surfaces are under consideration. This should enable the end user to better decide on which approach is most appropriate to their needs.

Both methods are applied to four different images shown that feature various damage forms, lighting conditions, viewing angles, resolutions, etc. These images are shown in Figure 6.14, along with the regions detected using the color based method and the texture analysis approach. The performances for each damage detection technique are presented in Table 6.2.

It may be noted from these results that the color based method was quite successful overall and it proved effective at locating the presence of damage as well as accurately defining the shape and size of damaged regions. The texture based method was effective at locating the presence of damage as may be observed from Figure 6.14, but it did not perform as well as the color based method at defining the extent of damage, which resulted in poorer DR, MCR, and δ values in Table 6.2. These results suggest the color based method has greater applicability for a wider range of damage forms.

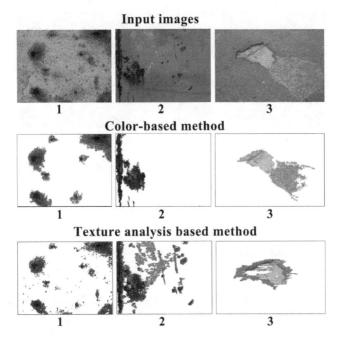

Figure 6.14 Top row: original images featuring damage: (1) pitting corrosion, (2) metal sheet with various corrosion damage, and (3) road delamination. Middle row: regions detected using the color based method. Bottom row: regions detected using texture analysis.

Table 6.2 Comparison of the described color based technique and texture analysis technique

Sample image	Color based method			Texture analysis		
	DR	MCR	δ	DR	MCR	δ
(1) Pitting corrosion	84%	8%	0.18	81%	22%	0.29
(2) Metal sheet with corrosion damage	90%	3%	0.11	96%	24%	0.24
(3) Road delamination	93%	10%	0.12	52%	10%	0.49

6.6 DISCUSSION AND CONCLUSION

Advances in computer systems have paved the way for the development of sophisticated damage detection techniques that can effectively capitalize on the ever-increasing level of computational efficiency available. The purpose of damage detection techniques is to locate and quantify the area occupied by visible damage (typically larger than 10^{-6} m^2) on the surface of infrastructural elements with minimal human supervision. In this chapter, two damage detection techniques were described. The first method utilized color information as the basis for damage detection, while the second method relied on texture information. The motivation for describing both color and texture based techniques stems from the fact that certain damage forms can only be effectively separated from the background based on either texture or color properties.

The color based method adopts a multi-phase segmentation methodology that incorporates features from three standard image processing and data analysis techniques. Since these techniques are well-known and described in the literature, the technique may be easily replicated and implemented. A key benefit of this technique is its ability to produce better defined and more homogeneous regions of interest without being affected by isolated extraneous pixels. This cleaner segmentation is achieved by efficiently integrating pixel and spatial relationships. The presented results indicate that the method can be successfully applied to a variety of damage forms and surface types.

The texture analysis technique involves generating texture related statistics for each pixel in the image. The pixels are consequently classified through support vector machines (SVM) models based on these texture features. Texture analysis based segmentation is usually more immune to variations in lighting conditions than color based techniques, where only the pixel intensity values are considered as opposed to texture-based segmentation techniques in which the relationship between adjacent pixel intensity values is considered. This relationship is often maintained to a significant extent even when inherent chromatic and luminous complexities are introduced to the scene.

The specific application presented in this chapter demonstrates the immediate success of the method as an NDT tool to assist visual inspections where an improved detection directly influences the owner of infrastructure systems during a decision-making process.

REFERENCES

Abdel-Qader, Ikhlas, Solange Yohali, Osama Abudayyeh, and Sherif Yehia. "Segmentation of thermal images for non-destructive evaluation of bridge decks." *Ndt & E International* 41, no. 5 (2008): 395–405.

Adams, Rolf, and Leanne Bischof. "Seeded region growing." *IEEE Transactions on Pattern Analysis and Machine Intelligence* 16, no. 6 (1994): 641–647.

Baraldi, Andrea, and Flavio Parmiggiani. "An investigation of the textural characteristics associated with gray level cooccurrence matrix statistical parameters." *IEEE Transactions on Geoscience and Remote Sensing* 33, no. 2 (1995): 293–304.

Chan, Tony F., and Luminita A. Vese. "Active contours without edges." *Image Processing, IEEE Transactions on* 10, no. 2 (2001): 266–277.

Cheddad, Abbas, Dzulkifli Mohamad, and Azizah Abd Manaf. "Exploiting Voronoi diagram properties in face segmentation and feature extraction." *Pattern Recognition* 41, no. 12 (2008): 3842–3859.

Choi, Byeongcheol, Seungwan Han, Byungho Chung, and Jaecheol Ryou. "Human body parts candidate segmentation using laws texture energy measures with skin color." In *Advanced Communication Technology (ICACT), 2011 13th International Conference on*, pp. 556–560. IEEE, 2011.

Cortes, Corinna, and Vladimir Vapnik. "Support-vector networks." *Machine Learning* 20, no. 3 (1995): 273–297.

Cristianini, Nello, and John Shawe-Taylor. *An introduction to support vector machines and other kernel-based learning methods.* Cambridge University Press, 2000.

Felzenszwalb, Pedro F., and Daniel P. Huttenlocher. "Efficient graph-based image segmentation." *International Journal of Computer Vision* 59, no. 2 (2004): 167–181.

Gadelmawla, E. S. "A vision system for surface roughness characterization using the gray level co-occurrence matrix." *NDT & e International* 37, no. 7 (2004): 577–588.

Gill, R. S. "Ice cover discrimination in the Greenland waters using first-order texture parameters of ERS SAR images." *International Journal of Remote Sensing* 20, no. 2 (1999): 373–385.

Giralt, Juan, L. Rodriguez-Benitez, J. Moreno-Garcia, C. Solana-Cipres, and L. Jimenez. "Lane mark segmentation and identification using statistical criteria on compressed video." *Integrated Computer-Aided Engineering* 20, no. 2 (2013): 143–155.

Haralick, Robert M., and Karthikeyan Shanmugam. "Textural features for image classification." *IEEE Transactions on Systems, Man, and Cybernetics* 6 (1973): 610–621.

Huang, Yishuo, and Jer-Wei Wu. "Infrared thermal image segmentations employing the multilayer level set method for non-destructive evaluation of layered structures." *NDT & E International* 43, no. 1 (2010): 34–44.

Kasban, H., O. Zahran, H. Arafa, M. El-Kordy, Sayed MS Elaraby, and FE Abd El-Samie. "Welding defect detection from radiography images with a cepstral approach." *NDT & E International* 44, no. 2 (2011): 226–231.

Li, Dawei, Lihong Xu, Erik D. Goodman, Yuan Xu, and Yang Wu. "Integrating a statistical background-foreground extraction algorithm and SVM classifier for pedestrian detection and tracking." *Integrated Computer-Aided Engineering* 20, no. 3 (2013): 201–216.

Liu, P. C. "Damage to concrete structures in a marine environment." *Materials and Structures* 24, no. 4 (1991): 302–307.

Lu, Chun S., Pau C. Chung, and Chih F. Chen. "Unsupervised texture segmentation via wavelet transform." *Pattern Recognition* 30, no. 5 (1997): 729–742.

Molero, Miguel, S. Aparicio, Ghaida Al-Assadi, M. J. Casati, M. G. Hernández, and J. J. Anaya. "Evaluation of freeze–thaw damage in concrete by ultrasonic imaging." *NDT & E International* 52 (2012): 86–94.

Naccari, Filippo, Sebastiano Battiato, Arcangelo Bruna, Alessandro Capra, and Alfio Castorina. "Natural scenes classification for color enhancement." *IEEE Transactions on Consumer Electronics* 51, no. 1 (2005): 234–239.

O'Byrne, Michael, Franck Schoefs, Bidisha Ghosh, and Vikram Pakrashi. "Texture analysis based damage detection of ageing infrastructural elements." *Computer-Aided Civil and Infrastructure Engineering* 28, no. 3 (2013): 162–177.

Otsu, Nobuyuki. "A threshold selection method from gray-level histograms." *IEEE Transactions on Systems, Man, and Cybernetics* 9, no. 1 (1979): 62–66.

Yazid, Haniza, H. Arof, Hafizal Yazid, Sahrim Ahmad, A. A. Mohamed, and F. Ahmad. "Discontinuities detection in welded joints based on inverse surface thresholding." *NDT & e International* 44, no. 7 (2011): 563–570.

Chapter 7

3D imaging

7.1 INTRODUCTION

Inspectors are continually faced with the question of how to maximize the quality of data obtained from inspection. Part of this question must consider how damage diagnostic tools can be fully utilized. So far in this book, we have seen that a camera is a powerful tool capable of capturing a rich amount of visual information from a scene that can subsequently be interpreted using damage detection algorithms. However, each photograph is still merely a 2D projection of the scene. By taking multiple photographs of a target from different vantage points, it is possible to encode 3D information in the imagery. Naturally, this 3D information lends itself to many applications, and it immediately enhances the level of information that can be obtained from inspection imagery. Access to 3D information enables engineers to extract physically meaningful measurements from reconstructed models of underwater targets, such as the size and shape of visible defects—detection of which can be automated using the techniques presented in the previous chapters.

This chapter looks at the main approaches for obtaining 3D information using camera systems. Following this, a step-by-step workflow for recovery 3D shape using stereo imaging is outlined. Special attention is given to key steps such as stereo camera calibration, finding matching points in two images, and generating a 3D mesh. The process is demonstrated on stereo images of a ship hull that is colonized by barnacles. Barnacle fouling is problematic as it reduces the hydrodynamic performance of marine vessels—so much so that the fuel consumption can increase by up to 40% to maintain a given speed (Chambers et al., 2006). Owners/operators of marine vessels, therefore, have a keen interest in monitoring the progression of biofouling so that they can choose the best times to carry out costly cleaning regimes, and so that they have more reliable estimates of fuel consumption levels.

7.2 APPROACHES FOR OBTAINING 3D INFORMATION

In recent decades, considerable advances have been made in developing underwater 3D shape recovery techniques involving coded structured light, structure from motion (SfM), and stereo photography. These advancements have been largely spearheaded by researchers in the fields of marine biology and underwater archaeology. Coded structured light approaches use a light projector to project a visible or non-visible (i.e., infrared) light pattern onto an object, as depicted in Figure 7.1(a). A camera (or multiple cameras) then captures how the pattern interacts with the scene. The downside of such approaches is that they are highly susceptible to the effects of underwater absorption and scattering, and as a result, they tend only to be effective over short ranges and are generally unable to resolve fine details (depending also on the density of the projected light pattern). Additionally, the projected light interferes with the scene and may partially mask detail on an object of interest. On a more practical note, these systems are not easily adaptable for underwater use, especially compared with camera-only systems.

Structure from motion (SfM) photogrammetry techniques are used to reconstruct a scene from a sequence of overlapping images acquired by a single moving camera (Hartley & Zisserman, 2003). The process is based on the automatic extraction of points of interest (a sparse set of features), the tracking of this sparse set of features across the image sequence, and the estimation of their 3D positions using multiple views (Cocito et al., 2003). SfM techniques have the advantage of being easy to implement on-site as they only require a diver to operate a single unconstraint camera and capture photographs as normal. However, they are limited insofar as (1) the 3D reconstructions are scale ambiguous (i.e., the 3D scene can only be recovered up to an unknown scale factor), (2) there is a requirement that the scene remains static, and (3) the performance of the technique is heavily reliant on the successful tracking of features over time. Reliably tracking

(a)　　　　　　　　　　　　　　**(b)**

Figure 7.1　(a) Illustration of coded structured light for obtaining depth information, and (b) remotely operated vehicles (ROVs) are typically equipped with at least one camera/video system that enables 3D shape to be recovered using structure from motion (SfM) or stereo photography.

features is difficult when operating in underwater conditions characterized by poor visibility (O'Byrne et al., 2014). Given these limitations, this chapter focuses on stereo systems as it is felt that the effort and expense associated with underwater inspections warrant the use of more dependable solutions.

Stereo systems are often used for underwater inspections, either carried by divers or attached to ROVs (Figure 7.1(b)). These systems consist of two horizontally displaced digital cameras to simultaneously photograph a scene from slightly different perspectives, resulting in two images that are collectively referred to as a stereo pair. 3D information can be extracted by comparing information about the scene from the vantage point of the left camera with that of the right camera. 3D shape information can be obtained from a single stereo pair once the stereo system has been calibrated, and unlike SfM, stereo systems can be used even if the scene does not remain static (i.e., it evolves over time) since both images in the stereo pair are captured simultaneously. One avenue that is increasingly being pursued by researchers, such as Xu et al. (2016), involves integrating stereo imagery into an SfM framework. This allows full 3D models (that show all accessible sides of the object) to be created and to be correctly scaled.

Other techniques for recovering 3D shape information include structure-from-shading, depth from defocus, and the use monocular cues to infer 3D shape; however, these approaches are often crude and are ill-suited for accurate quantitative shape recovery.

7.3 STEREO IMAGING

The workflow for creating 3D reconstructions using stereo imaging is shown in Figure 7.2.

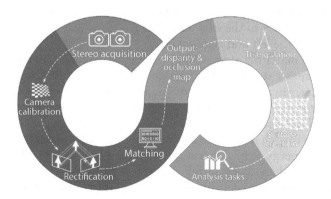

Figure 7.2 **Pipeline for 3D reconstruction using stereo imaging.**

Left stereo image　　　　**Right stereo image**

Figure 7.3 Left and right images of a stereo pair. These images are of a barnacle-covered surface.

The first step involves acquiring the stereo imagery. Details on how to properly configure a stereo system are covered in Section 3.3.2. This demonstration only considers one stereo pair, as shown in Figure 7.3, which consists of a left and right image of a barnacle-covered surface. These images have a resolution of 1600 × 900 pixels.

The next step is camera calibration. Calibration is needed to ensure a Euclidean reconstruction, that is, a reconstruction where the 3D model is appropriately scaled, and angles remain true to reality. This demonstration focuses on checkerboard-based calibration; however, strategies for performing self-calibration are also discussed. Self-calibration, or autocalibration, is the process of obtaining intrinsic camera parameters (such as the focal length) and extrinsic parameters (the rotation and translation transformations between stereo cameras) using only the constraints in the scene (Zhang et al., 1996).

Camera calibration feeds into the next step, which is rectification. Rectification entails transforming the stereo images such that corresponding points in the left and right images are separated only by a horizontal offset and not by a vertical offset. Rectification significantly decreases the computational complexity of finding corresponding points and it is an important precursor to the stereo matching step, which seeks to match the same points in the left and right images.

The stereo matching step is a key part of the 3D reconstruction pipeline. While many stereo matching algorithms have been devised to solve this problem, they often perform poorly when applied to underwater images due to the poor visibility conditions and the complex underwater light field. As such, an advanced stereo correspondence algorithm is described in this chapter. The output from this algorithm is a disparity map and an energy map. The energy map identifies points where disparity values are likely to be unreliable. With knowledge of the disparity values and the camera parameters, a 3D point cloud can be generated via triangulation.

A good-to-know follow-up step involves surface reconstruction, whereby the 3D point cloud is upgraded to a watertight mesh. This step may be of value to engineers looking to carry out further analysis using the reconstructed 3D models since a polygon mesh is the expected input for many upstream engineering software packages such as fluid-structure interaction (FSI) programs.

7.3.1 Calibration

To make visual data a part of a quantitative assessment, it is necessary to calibrate the stereo camera system so that pixels in the image can be related to the 3D points in the scene. Camera calibration involves finding the camera's extrinsic and intrinsic parameters. The extrinsic parameters describe the rotation and a translation of the right camera with respect to the left camera, while the intrinsic parameters consist of the focal length, the optical center, also known as the principal point, and the skew coefficient. Calibration can also correct for lens distortion.

To understand the role that the intrinsic and extrinsic parameters play for recovering 3D shape, it is helpful to look at the idealized pinhole camera model shown in Figure 7.4. In this widely used model, the relationships between world coordinates $\mathbf{X}$ and image (pixel) coordinates $\mathbf{x}$ is modelled via the perspective transformation.

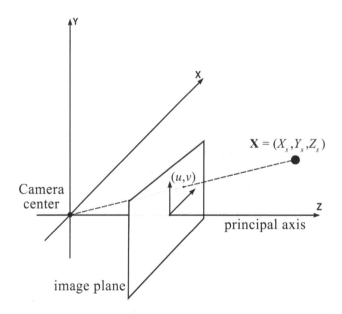

Figure 7.4 Pinhole camera model that shows the relationship between a world point, X, and a point that is projected onto the image plane that is denoted by (u, v).

Let $\mathbf{x} = (u, v, 1)$ be the homogeneous coordinates of a point in an image. The equation for relating these image points to real-world points is:

$$s\begin{bmatrix} u \\ v \\ 1 \end{bmatrix} = \mathbf{P} \begin{bmatrix} X_s \\ Y_s \\ Z_s \\ 1 \end{bmatrix} \tag{7.1}$$

where s is a non-zero scale factor, $\mathbf{X} = (X_s, Y_s, Z_s, 1)$ are world coordinates, and $\mathbf{P}$ is a 3×4 projection matrix that completely represents the mapping from the scene to the image. $\mathbf{P}$ encapsulates both the extrinsic and intrinsic parameters of a camera. It is given by

$$\mathbf{P} = \mathbf{K}\begin{bmatrix} \mathbf{R} | \mathbf{T} \end{bmatrix} \tag{7.2}$$

where $\mathbf{R}$ and $\mathbf{T}$ are the extrinsic parameters representing the rotation matrix and the translation of the camera in the 3D scene, respectively. Both the rotation matrix and the translation vector have three degrees of freedom. $\mathbf{K}$ is an upper-triangular matrix that represents the intrinsic parameters. It has five degrees of freedom. The five intrinsic parameters correspond to the focal length in pixels for the horizontal and vertical axes, f_x, and f_y, respectively, the skew parameter, sk, and the two principal points x_0 and y_0 (the optical centre of the camera), as shown in Equation 7.3.

$$\mathbf{K} = \begin{bmatrix} f_x & sk & x_0 \\ 0 & f_y & y_0 \\ 0 & 0 & 1 \end{bmatrix} \tag{7.3}$$

Thus, an image point can be related to the corresponding point in the 3D world space with knowledge of the extrinsic and intrinsic parameters, and the scale factor s, which may be determined using an object of known dimension in the scene (such as the size of a square on the calibration checkerboard) or if the baseline distance between the left and right cameras is known.

7.3.1.1 Checkerboard-based calibration

Checkerboard-based calibration entails photographing a high-contrast checkerboard pattern from several angles, ensuring that the entire checkerboard is visible by both the left and right camera frames. In the case of videoing, the checkerboard should be recorded from different sides for a short while (i.e., about a minute). Individual frames can then be extracted from

Left calibration images   **Right calibration images**

Figure 7.5 Left and right stereo images of a checkerboard calibration pattern.

the video and processed in the same manner as still photographs. A minimum of two calibration images of the checkerboard per camera is required; however, it is advisable to capture as many as possible from a range of angles and distances (e.g., at least 10 images per camera). This helps to obtain more accurate camera parameters. Having many calibration images is especially important when using wide-angle cameras where a high degree of lens distortion is normally present. In this example, for ease of demonstration, only two calibration images per camera are captured. These are shown in Figure 7.5.

Calibration is quite a straightforward process in MATLAB®. The code for reading in the calibration images and automatically detecting the corners of the checkerboard is as follows:

```
1  % Specify filenames of the calibration images to process
2  LeftImages = {'Left_calib_0001.tif','Left_calib_0002.
   tif'};
3  RightImages = {'Right_calib_0001.tif','Right_calib_0002.
   tif'};
4
5  % Automatically detect checkerboards in images
6  [imPoints, boardSize, imagesUsed] =...
7      detectCheckerboardPoints(LeftImages, RightImages);
8
9  % Show detected corners in the left images
10   for i = 1:numel(LeftImages)
11       I = imread(LeftImages{i});
12       subplot(1, 2, i); imshow(I);
13       hold on; plot(imPoints(:,1,i), imPoints(:,2,i),
         'o');
14   end
15
16  % Generate world coordinates of the checkerboard
    keypoints
17  squareSize = 25; % Specify size in real-world units of
    'mm'
```

```
18 worldPoints = generateCheckerboardPoints(boardSize,
19 squareSize);
10
21 % Calibrate the camera
22 [stereoParams, pairsUsed, Errors] =...
23 estimateCameraParameters(imPoints, worldPoints,...
24 'EstimateSkew', false,'EstimateTangentialDistortion',fa
   lse,...
25 'NumRadialDistortionCoefficients', 2, 'WorldUnits',
   'mm',...
26 'InitialIntrinsicMatrix', [], 'InitialRadialDistortion',
   []);
27
28 % Visualise pattern locations
29 figure, showExtrinsics(stereoParams, 'CameraCentric');
```

The corner points of the planar checkerboard have now been detected, as shown in Figure 7.6.

This code sample also outputs the pose of the checkerboards in 3D space (Figure 7.7) and, more importantly, the pose of the two stereo cameras (which are defined by the extrinsic parameters).

The estimated camera parameters are summarized in Table 7.1.

Calibration is valid as long as there are no consequential changes made to the stereo system or the operating environment. The intrinsic parameters will change if:

- Switching between camera and video modes
- Changing the resolution of the imagery
- Changing the field of view/focal length
- Transitioning between underwater filming and above water filming (this is because the refraction means the focal length is effectively 25% longer for underwater filming). Therefore, calibration should be carried out in an underwater medium if the inspection imagery is also captured underwater.

Figure 7.6 Detected corners in the calibration images from the left camera.

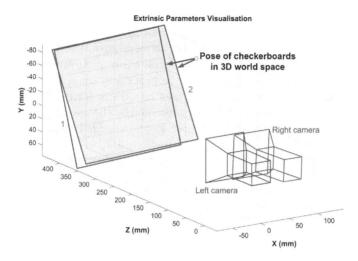

Figure 7.7 Schematic of the stereo cameras and the pose of the checkerboard patterns in 3D world space.

Table 7.1 Intrinsic and extrinsic parameters of the stereo system

Intrinsic parameters for the left camera	$\mathbf{K}_{\text{left}} = \begin{bmatrix} f_x = 1555 & sk = 0 & x_0 = 800 \\ 0 & f_y = 1555 & y_0 = 451 \\ 0 & 0 & 1 \end{bmatrix}$
Intrinsic parameters for the left camera	$\mathbf{K}_{\text{right}} = \begin{bmatrix} f_x = 1555 & sk = 0 & x_0 = 800 \\ 0 & f_y = 1556 & y_0 = 451 \\ 0 & 0 & 1 \end{bmatrix}$
Rotation of the right camera with respect to the left camera	$\mathbf{R} = \begin{bmatrix} 0.994 & -0.042 & -0.102 \\ 0.042 & 0.999 & -0.005 \\ 0.102 & 0.001 & 0.995 \end{bmatrix}$
Position of the right camera with respect to the left camera	$\mathbf{T} = \begin{bmatrix} -53.498 & -0.448 & 4.01 \end{bmatrix}$

The extrinsic parameters will change when the pose (i.e., the position and orientation) of one camera changes with respect to the other camera. Therefore, it is important that the cameras are securely fixed to a rigid base tray throughout the duration of inspection so that the calibration parameters remain relevant. It is usually a good idea to configure the stereo system on-site and perform camera calibration prior to collecting the inspection imagery.

7.3.1.2 Self-calibration

Self-calibration, also known as auto-calibration, is an attractive way of determining the intrinsic and extrinsic camera parameters as it does not require the diver/photographer to undertake any preliminary calibration procedures. Self-calibration refers to the process of obtaining a calibrated camera matrix using the static scene as a constraint for the five degree-of-freedom pinhole camera model, as represented by the matrix K in Equation 7.3. Theoretically, a minimum of three views is needed for full calibration assuming the intrinsic parameters remain constant between views (i.e., the focal length, or any other parameter, is not adjusted between views). In reality, however, the principal points, x_0 and y_0, can usually be safely estimated to be at the center of the image. This is the reflected in the intrinsic parameters that were found through checkerboard calibration (Table 7.1)—x_0 and y_0 are approximately half the image width (1600 pixels) and half the image height (900 pixels). Furthermore, most modern imaging sensors and optics provide further prior constraints such as zero skew and unity aspect ratio, which means that the horizontal and vertical focal lengths, f_x, and f_y, can be assumed to have the same value. Integrating these priors will reduce the minimum number of views required to two. The only other information that is required is the size of an object in the scene or the baseline distance. This piece of information is necessary for obtaining correctly scaled 3D reconstruction.

There are some inherent drawbacks associated with self-calibration. First, real cameras are affected by radial and tangential lens distortion. While checkerboard calibration is capable of estimating the distortion parameters and undistorting the images, self-calibration methods are not well equipped to deal with this additional layer of complexity. This is especially problematic when dealing with wide-angle lenses as the level of distortion tends to be particularly severe.

Second, self-calibration requires the extraction of a sparse set of corresponding points from the left and right frames of the stereo pair, x and x', respectively, in order to set up the static scene constraint problem. However, if not enough corresponding points are found (a minimum of 5 is required), or if erroneous correspondences are present, then the essential matrix cannot be determined, or it will be poorly estimated. This issue is mitigated by using the robust, state-of-the-art SIFT (Scale Invariant Feature Transform) algorithm (Lowe, 1999) for extracting and matching points of interest in the left and right images in conjunction with bundle adjustment, which serves to reject outlier matches.

Finally, self-calibration is quite complicated to implement programmatically. Non-linear optimization is often used to improve the results; however, there is no guarantee of convergence. Moreover, the results are rarely as good as those obtained through conventional checkerboard-based calibration (O'Byrne et al., 2016). Nevertheless, interested readers are directed toward the works of Faugeras et al. (1992), where the self-calibration process is described in detail.

A crude estimate of the intrinsic parameters (for a stereo system where the left and right cameras are the same) can be found by estimating the focal lengths, f_x and f_y, in pixels and using the default/assumed values for all of the other parameters. The focal length in pixels can be approximated using the formula given in Equation 7.4.

$$\text{focal length in pixels} \approx \frac{\lambda \times (\text{image width in pixels}) \times (\text{focal length in mm})}{(\text{sensor width in mm})}$$

$$(7.4)$$

where λ accounts for the medium—if the imagery is acquired above water then λ has a value of 1, however, then the effective focal length is around 25% greater underwater so therefore λ is assigned a value of 1.25 when dealing with underwater image collection. The dimensions of a camera's sensor can typically be found online. Common sensor sizes are provided in Section 3.2.1. As an example, for a full frame sensor (36 × 24 mm) that records video at a resolution of 1600 × 900 pixels, using a lens with a focal length of 35 mm, and collecting imagery underwater ($\lambda = 1.25$), an estimate for the focal length in pixels is given by:

$$f_x = f_y \approx \frac{1.25 \times (1600 \text{ pixels}) \times (28 \text{ mm})}{(36 \text{ mm})}$$
$$f_x = f_y \approx 1556 \text{ pixels}$$

$$(7.5)$$

The estimated camera calibration matrices for the left and right cameras would then be:

$$\mathbf{K}_{\text{left}} = \mathbf{K}_{\text{right}} = \begin{bmatrix} f_x = 1556 & sk = 0 & x_0 = 800 \\ 0 & f_y = 1556 & y_0 = 450 \\ 0 & 0 & 1 \end{bmatrix}$$

$$(7.6)$$

7.3.2 Rectification

Rectification is the process of transforming the left and right images of a stereo pair, such that corresponding points in each image are separated only by a horizontal distance and not by a vertical distance. Rectification can be carried out either with calibration information using the essential matrix or without it using the fundamental matrix. The essential matrix, $\mathbf{E}$, is a 3×3 matrix that depends only on the extrinsic parameters **R** and **T**. The fundamental matrix, **F**, is a generalization of the essential matrix. It relates corresponding points in the stereo images and may be estimated from at least seven-point correspondences. The seven parameters represent the only geometric information about cameras that can be obtained through point correspondences alone. It does not require any knowledge of camera internal parameters.

All points in the rectified images should satisfy the epipolar geometry of a rectified image pair (i.e., that the images are aligned horizontally). This may be expressed as follows: if a point x'_i in the left image corresponds to a point x'_i in the right image, then they should satisfy the constraints in Equation 7.7 and 7.8. These constraints are geometrically equivalent, the only difference being that Equation 7.7 is more general as it is based on the fundamental matrix, **F**, while Equation 7.8 makes use of camera calibration information by including the calibration matrices for the left and right cameras, **K** and **K'**, respectively, as well as the relative position and orientation between the cameras, which is captured by the essential matrix **E**:

$$x'^T_i \mathbf{F} x_i = 0 \tag{7.7}$$

$$x'^T_i \mathbf{K'} x_i = 0^{-T} \mathbf{E} \mathbf{K}^{-1} x_i = 0 \tag{7.8}$$

Rectification makes the task of matching pixels in the left and right image considerably faster as it confines the search space to the horizontal axis only. Images can be rectified using the calibration information that was established in the previous section, as per the following code:

```
1 % Read in left and right images (shown in Figure 7.3).
2 LeftIM = imread('LeftImage.tif');
3 RightIM = imread('RightImage.tif');
4
5 % Perform rectification on these images using the
  calibration
6 information stored in 'stereoParams'
7 [Leftr, Rightr] = rectifyStereoImages(LeftIM,
8 RightIM,...stereoParams,'OutputView','valid');
9
10 % Display the results
11 figure; imshowpair(Leftr, Rightr,'montage');
```

The rectified left and right images are shown in Figure 7.8. The images have been transformed in such a way that every point in the left image is now separated only by a horizontal offset from the corresponding point in the right image, and not by any vertical offset.

| Left rectified image | Right rectified image |

Figure 7.8 Rectified left and right images.

7.3.3 Stereo correspondence algorithm

The next stage of the stereo process involves finding matching points between the left and right stereo images. This task can be quite challenging due to a host of reasons. Issues such as specular reflections (bright spots) in the scene, textureless surfaces, noisy/grainy images, blurry images, large perspective differences between left and right cameras, etc. give rise to false matches. A wide variety of stereo matching algorithms have been developed—a detailed taxonomy is provided by Olofsson (2010) and Scharstein and Szeliski (2002). Matching algorithms may be broadly classified into two categories: local and global methods.

Local methods are generally based on computing the matching cost between small patches, or windows, in the reference image (normally the left camera image) with potential matching patches in the other stereo image pair, followed by aggregation of the computed costs. The disparity value that is assigned to each point in the reference image is the disparity associated with the minimum matching cost. Local methods are capable of running in real-time, making them suitable for a range of online applications such as for automated inventorying of road signs (Wang et al., 2010) and traffic surveillance (Sappa et al., 2008). However, their computational efficiency comes at the expense of reduced matching accuracy and increased sensitivity to noise.

For underwater inspections, stereo images can be processed off-line, and as such, there is no requirement for real-time processing. Instead, the emphasis lies in obtaining more reliable disparity measurements, which warrants the use of global matching methods. In comparison to local methods, global methods demonstrate better handling of texture-less regions and better tolerance to noise. This results in more accurate disparity maps. Global methods generally entail computing the matching cost followed by optimizing the disparity map. Disparity optimization is usually done by minimizing some energy function, usually consisting of a correspondence data term and a smoothness term.

The smoothness term penalises cases where adjacent pixels have different disparity values.

Popular types of global methods include dynamic programming (Zhuang and Wang, 2009) and Markov random field (MRF) based methods. The latter have attracted a lot of recent interest owing to their strong performances in the realm of stereo. Moreover, the emergence of fast algorithms for approximate inference, such as Graph Cuts (Papadakis & Caselles, 2010) and Belief Propagation (BP) (Sun et al., 2002), have enabled MRF-based methods to become computationally viable. These methods have been found to yield good results thanks in a large part to how they explicitly model smoothness. For this reason, a BP-based technique is described and implemented in this chapter. The output of this stereo matching technique is a dense disparity map that represents the distance in pixels between corresponding points in the rectified images.

7.3.3.1 Matching cost computation

Given two images of the same scene, the stereo matching algorithm aims to match pixels in one image with the corresponding pixels in the other image. A similarity measure is used to evaluate how closely a patch in the reference image (normally the left image) resembles a candidate patch in the other stereo image, and thus how likely the patches depict the same region in the scene. Patches may be compared on the basis of pixel intensity values, texture patterns, or using census/rank-transformed data. Common measures of similarity used for computing the matching cost include sum of squared distances (SSD), sum of absolute distances, normalized-cross correlation, and Hamming distance (Giachetti, 2000). There are advantages and drawbacks associated with each similarity measure. Correlation-based metrics are capable of tolerating variations in image brightness, making them a good choice for situations where challenging lighting conditions exist, or when the left and right cameras in a stereo system have different exposure settings. Likewise, census transform-based matching is known to be robust to radiometric distortion since global lighting differences between two images will not affect the ordering of pixels at a local level. The census transform summarizes local image structure. It is based on the relative ordering of local intensity values and not on the intensity values themselves. This aspect means that it can tolerate outliers and perform better near object boundaries compared with only using color data (Tavera-Vaca et al., 2015). However, these matching metrics often take a longer time to compute compared with simple similarity measure such as the sum of squared distances (SSD), which is the metric used in this demonstration.

Calculating the SSD score between a point in the left image and points in the right image involves the use of a sliding window. The window moves horizontally over the predefined disparity search range in the right image. At every disparity value, the window in the right image is compared with

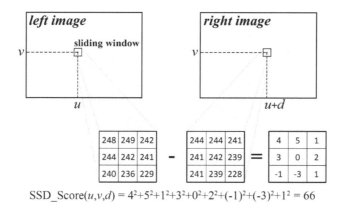

SSD_Score$(u,v,d) = 4^2+5^2+1^2+3^2+0^2+2^2+(-1)^2+(-3)^2+1^2 = 66$

Figure 7.9 Example of computing the sum of squared difference (SSD) score between a patch in the left image with a patch in the right image.

the reference window in the left image. The SSD score has a small value when there is a high degree of similarity between the two windows, and conversely, it has a large value when the two windows are dissimilar. The SSD score associated with matching a point (u,v) in the left image to a point $(u+d,v)$ in the right image is illustrated for a 3×3 size window in Figure 7.9.

A window size of 9×9 is used in this demonstration. It represents a reasonable balance between capturing the intensity variation within a window while keeping the window specific to the point on which it is centered upon. Strategies for choosing an optimal window size are discussed by Kanade and Okutomi (1994).

The following code shows how to compute the matching cost using SSD as a matching metric. The inputs are a rectified stereo pair and the disparity range (the disparity levels over which to carry out the search for matching pixels).

```
1 labels = 15;% No. of disparity labels
2 d_min = 140;% Smallest disparity value
3 Direction = [1,2,3,4,5];% Directions to pass messages
4 W =4;% window radius. The window size is (2w+1)x(2w+1)
  =9x9
5
6 [M,N,~] = size(Leftr);
7 d_max = d_min + labels-1;
8
9 % Pad the image so search does not extend beyond image
  boundary
10 pmax = max(abs(d_max),abs(d_min));
```

```
11 padLeft = (padarray((Leftr),[w w+pmax],'replicate'));
12 padRight = (padarray((Rightr),[w w+pmax],'replicate'));
13
14 mrf = double(zeros(M,N,length(Direction),labels));% set
   up MRF
15
16 Score = zeros(1,labels);% Initiate the matching score
   array
17
18 for y = 1+w:M-w
19
20     for x = 1+pmax+w:N-w-pmax
21
22     for d = d_min:d_max
23       % Compute the SSD at every disparity level
24         Score(d-d_min+1) = sum(sum(sum((padLeft(y-w:y+w,x-
25 w:x+w,:)...
26             - padRight(y-w:y+w,x-w+d:x+w+d,:)).^2)));
27     end
28
29     mrf(y,x,1,1:labels) =  Score;% Pass raw scores to
       the MRF
30     end
31 end
32
```

At this point, the disparity map could simply be created by choosing the disparity label that corresponds to the lowest SSD score. However, this would inevitably produce a noisy disparity map with many inaccurate disparities. Instead, we can exploit the fact that pixels in close proximity to one another generally share similar disparity values, unless the pixels represent an edge boundary in the scene. Therefore, the quality of the disparity map can be improved by utilizing this information.

7.3.3.2 Belief propagation on a Markov random field

This section shows how the problem of disparity map estimation can be formulated on an MRF and solved using BP. An MRF is a graphical model of a joint probability distribution consisting of visible and hidden nodes. Visible nodes are present at each pixel location and they represent information that is known or that can be computed (e.g., pixel color). Hidden nodes represent unknown values (e.g., disparity). Statistical dependencies between hidden variables are expressed by explicitly associating hidden variables with one another. MRFs have application in many image-processing tasks, including segmentation (Nguyen & Wu, 2013), fusion (Sun et al., 2013), and stereo matching (Sun et al., 2003). For stereo matching purposes, MRFs enable spatial dependencies between nearby pixels to be encoded. MRFs are comprised of nodes and links and may have loops (i.e., they may be cyclic), which is why belief propagation algorithms applied on these networks are sometimes referred to as loopy belief propagation algorithms. The presence of loops means there is no guarantee of convergence to the optimal global solution regardless of the number of iterations of the belief propagation algorithm. Despite this, belief propagation approaches are known to yield good approximate solutions. An example of how MRFs can be used to model the stereo problem is presented in Figure 7.10. $B = \{b\}$ represents the observed nodes (the matching costs), and $A = \{a\}$ represents the hidden nodes (the unknown disparity labels).

For simplicity, pixel locations in the image are denoted by a single index, f, as opposed to distinct horizontal and vertical spatial indices, (u,v), as before. It follows that f may be expressed by $f = u + (v - 1) \times M$ where M is the image width. For every pixel in the image, there exists an observed node b_f that represents the matching cost (as computed in the previous section), and a corresponding hidden node, a_f, which represents the unknown disparity value. The observed nodes are white and the hidden nodes are gray in Figure 7.9. The links between each node indicate a dependency. It may be noted that while the hidden variables are connected to their neighbors, the observed data points are connected only to a single hidden variable. As such, the estimated disparity at a point depends on the estimated disparity values of the four immediately adjacent pixels as well as the observed matching cost.

An optimal disparity map with minimum energy is approximated using loopy belief propagation where the message passing takes place between adjacent segments. There are various kinds of BP algorithms with different message update rules: max-product, sum-product, and min-sum (Szeliski et al., 2006). In this demonstration, min-sum is used where the objective is

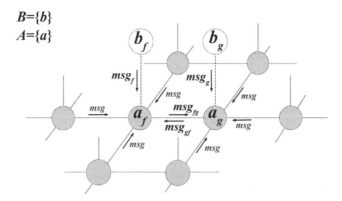

Figure 7.10 **Message passing on an MRF. Grey nodes are hidden variables. White nodes are observable variables.**

to minimize the energy function consisting of a matching cost term and a smoothness term:

$$E(B, A) = \sum_f C_f(p_f) + \sum_{(f,g) \in L} S(p_f, p_g) \qquad (7.9)$$

where E is the energy associated with having a disparity label p_f at every node a_f. $Cf(p_f)$ gives the cost of assigning a disparity label p_f at the point f in the image, as found from Equation 2. The second term, S, is the pairwise term, or the smoothness term, which penalizes instances of neighboring pixels f and g having different disparity labels, where g is the spatial index of a point that is directly adjacent to f. The smoothness term is based on the difference in gray level between neighboring pixels (or the gradient magnitude). In places where there is a large difference between adjacent pixels (which is often indicative of an edge boundary), the smoothness term is small so as to not overly enforce the smoothness constraint when there may be a genuine depth discontinuity in the scene. For small differences between adjacent pixels, the smoothness term has a large value so as to promote consistent and steady labelling across neighboring hidden nodes. The smoothness cost is low when neighboring pixels are assigned similar disparity values (i.e., p_f and p_g are similar). On a related note, using a smoothness term like this one has the advantage of being eligible for expedited BP computation, as outlined in the following MATLAB® code:

```
1  %% Belief Propagation on a Markov Random Field
2  msgL = zeros(M,N,5,labels);
3
4  %  Compute differences between adjacent pixels (i.e., a
5  gradient% map). This controls the smoothness term
6  LeftGray = rgb2gray(Leftr);
7  diff = (padarray(LeftGray(2:M-1,2:N-1),[1
   1],'replicate'));
8  % North, South, East and West differences
9  deltaN = abs(diff(1:M-2,2:N-1) - LeftGray(2:M-1,2:N-1));
10 deltaS =  abs(diff(3:M,2:N-1) - LeftGray(2:M-1,2:N-1));
11 deltaE =  abs(diff(2:M-1,3:N) - LeftGray(2:M-1,2:N-1));
12 deltaW =  abs(diff(2:M-1,1:N-2)
    - LeftGray(2:M-1,2:N-1));
13
14 Carry out 20 iterations of the BP algorithm
15 for tt = 1:20
16 % Update belief of every disparity label at all points
    in image
17 for Dir = 1:5; msgL(:,:,Dir,:) =
18 sum(mrf(:,:,Direction(Direction~=Dir),:),3);
```

```
19 end
20
21 for Dir = 1:5
22 C1 = 0.2; C2 = 20;% Adjustable parameters
23 % Use different gradient map depending on message
   direction
24 if Dir == 1; tmpEdge = zeros(M, N);
25 elseif Dir == 3; tmpEdge(2:M-1,2:N-1) =
   C1.*exp(-C2.*(deltaE));
26 elseif Dir == 2; tmpEdge(2:M-1,2:N-1) =
   C1.*exp(-C2.*(deltaW));
27 elseif Dir == 4; tmpEdge(2:M-1,2:N-1) =
   C1.*exp(-C2.*(deltaN));
28 elseif Dir == 5; tmpEdge(2:M-1,2:N-1) =
   C1.*exp(-C2.*(deltaS));
29 end
30
31 % Shape of cost function that penalizes adjacent points
   with%
32 disparate labels
33 paircost = (1-[0.5 0.25 0 0.25 0.5]).^2;
34
35 for i = 1:labels; counterk =0;
36
37 for j = i-floor(5/2):i+floor(5/2); counterk =
   counterk+1;
38 if j > 0 && j<=labels; msgL(:,:,Dir,j) =
   msgL(:,:,Dir,j) -
39 paircost(counterk).*tmpEdge;
40 end;end
41
42 new_msgL(:,:,Dir,i)=min(msgL(:,:,Dir,:) ,[],4)+tmpEdge;
43
44 counterk =0;
45 for j = i-floor(5/2):i+floor(5/2); counterk =
   counterk+1;
46 if j > 0 && j<=labels; msgL(:,:,Dir,j) =
   msgL(:,:,Dir,j) +
47  paircost(counterk).*tmpEdge;
48 end; end;
49 end
50 end
51
52    y = 2:M-1;
53    x = 2:N-1;
54    % Update the MRF
55    for Dir = Direction(1:end)
56    if Dir == 2; mrf(y,x-1,3,:) = new_msgL(y,x,Dir,:);
57    elseif Dir == 3; mrf(y,x+1,2,:) =
      new_msgL(y,x,Dir,:);
```

```
58      elseif Dir == 4; mrf(y-1,x,5,:) =
        new_msgL(y,x,Dir,:);
59      elseif Dir == 5; mrf(y+1,x,4,:) =
        new_msgL(y,x,Dir,:);
60      end; end
61 end
```

After 20 iterations of the BP algorithm, the best disparity label at every pixel can be found by calculating the belief vector bel_f, which is defined over each label p by Equation 7.10:

$$bel_f(p) = C_f(p) + \sum_{g \in J(f)} msg_{gf}^\phi(p) \tag{7.10}$$

The label p_f corresponding the smallest component of bel_f is taken as the MAP solution for the node a_f. As with most iterative algorithms, it can be run for a fixed number of iterations or terminated when the change in energy drops below a certain threshold. After 20 iterations, however, the system is generally stable and there tends to be only minimal gains in accuracy which often do not warrant the increased computational effort. The label p_f is assigned to the disparity map D at the point f, i.e., $D_f = p_f$.

```
1 %% Finds the MAP assignment as well as calculating the
  cost map
2 depthMap = zeros(M,N);% Initiate output depth map
3 costMap = zeros(M,N);% Initiate output cost map
4
5 for i = 1:M
6 for j = 1:N
7
8 costs(1:labels) = sum(mrf(i,j,1:5,1:labels),3);
9
10 % Interpolate discrete labels to obtain sub-pixel
   disparity
11 cs = spline(1:labels,[0 costs 0]);
12 xx = linspace(1,labels,101);
13
14 % Find the minimum cost and the corresponding disparity
   value
15 [ys,ind] = min(ppval(cs,xx));
16
17 % Populate the depth map and cost map with these values
18 depthMap(i,j) = d_min-1+xx(ind(1));
19 costMap(i,j) = ys(1);
20 end
21 end
```

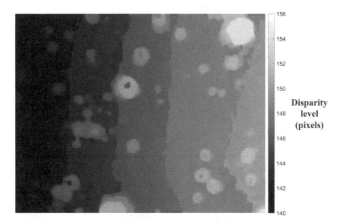

Figure 7.11 Output disparity map following the stereo matching phase. The disparity value at every point represents the horizontal distance between that point and the corresponding point in the other stereo image.

The output disparity map from this matching phase is shown in Figure 7.11. The disparity map produced by this method when applied to a real underwater stereo image pair is shown in Figure 7.12. This stereo image pair is taken from the repository that is described in Chapter 8. The disparity maps of two other stereo matching methods, namely a semi-global

Figure 7.12 Right stereo image and the disparity maps produced by three stereo matching methods.

matching method (Alagoz, 2008) and a basic block matching method (Birchfield & Tomasi, 1998), are shown for comparison purposes.

Having found corresponding points in the left and right camera images, the next step is to create a 3D point cloud via triangulation.

7.4 TRIANGULATION

Once a point in the left and right images has been successfully matched, it is possible to compute the location of the point in 3D world space by finding the intersection of the rays passing through the left camera's center of projection and $x^l = (u, v)$, and passing through the right camera's center of projection and $x^r = (u + d_{uv}, v)$. Here, u and v indicate the horizontal and vertical spatial coordinates of a pixel, respectively, and d_{uv} is the disparity value at (u,v), as found in the previous stage. This process is known as triangulation. With reference to Figure 7.13, a point X in the 3D world space lies at the intersection of rays projected from the centre of each camera and through the point on the image plane.

Triangulation can be performed with the following MATLAB® code:

```
1 % Reconstruct 3D scene.
2 points3D = reconstructScene(depthMap, stereoParams);
3 points3D = points3D./1000;% Convert from millimeters to
  meters
4
5 % Create a point color and use the left rectified
6 % color image to color the vertices
```

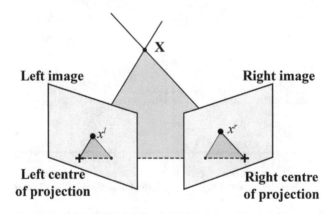

Figure 7.13 Projection of a 3D point X onto points in the left and right images.

3D Point Cloud

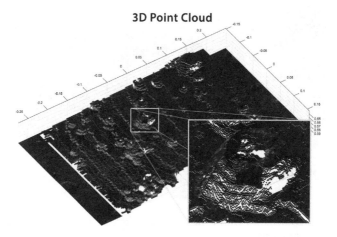

Figure 7.14 **3D point cloud.**

```
7 ptCloud = pointCloud(points3D, 'Color', im2uint8(Leftr));
8 pcshow(ptCloud)
9
10 % Save the point cloud in PLY format
11 pcwrite(ptCloud,'BarnacleSurface.ply')
12
```

The resulting point cloud is presented in Figure 7.14.

7.5 SURFACE RECONSTRUCTION

The dense 3D point cloud can be quite difficult to work with, especially if there are a lot of noisy or extraneous points, and are generally not directly usable in most 3D applications. For this reason, it is often desired to convert the point cloud to a polygon mesh model through a process commonly referred to as surface reconstruction. There are many techniques for converting a point cloud to a 3D surface. This section shows one approach whereby the 3D point cloud generated in the previous section is imported into Meshlab and a polygon mesh is constructed using the Poisson surface reconstruction algorithm (Kazhdan & Hoppe, 2013). MeshLab is an open source tool for managing point clouds and converting them into 3D triangular meshes.

The first step is to open the 3D point cloud in Meshlab. In this example, the 3D point cloud was saved as a PLY file named 'BarnacleSurface.ply' in the previous section. Once opened, the user can interactively select and

Figure 7.15 **Example showing how measurements can be extracted from the reconstructed mesh.**

remove extensors pixels. Once the 3D point cloud has been pruned, the normals for the point set can then be computed. Normals are important properties of a geometric surface and are used for a variety of reasons such as for correctly illuminating a surface. Given a geometric surface, it's usually straightforward to establish the direction of the normal at a certain point on the surface as the vector perpendicular to the surface at that point. However, since the 3D point clouds only represent a set of point samples, we must compute the normal at every point by looking at neighboring points and estimating the underlying surface. This task can be completed by navigating to *Filters>Point Set>Compute normals for point set*. There is the option to adjust certain parameters such as the number of neighboring points to consider. The default parameter values should be adequate, however.

The next step is Poisson surface reconstruction. This task can be completed by navigating to *Filters>Remeshing, Simplification and Reconstruction> Screened Poisson Surface Reconstruction*. This surface reconstruction algorithm creates watertight surfaces from orientated point sets (i.e., point sets where the normals have been estimated). The reconstructed surface is shown in Figure 7.15, where it is possible to inspect the mesh and extract measurements.

7.6 SUMMARY

Recovering 3D shape information is a challenging but useful task and has wide applicability in many areas of Structural Health Monitoring (SHM). This chapter describes and demonstrates the complete workflow for extracting 3D information using stereo imaging. The first major step that is covered is camera calibration. Determining the intrinsic and extrinsic camera parameters is crucial for producing an accurate 3D reconstruction that can be expressed in terms of real-world units, such as millimeters. This task is often overlooked despite being an integral part of the stereo imaging pipeline.

The next key step that is thoroughly addressed is how to solve the stereo correspondence problem, that is, the problem of finding the same points in the left and right camera images. This problem is non-trivial when the imagery is collected in challenging visibility conditions characterized by non-uniform lighting and limited range visibility. A robust stereo matching algorithm based on belief propagation (BP) is described, which is well-suited for poor visibility conditions. With knowledge of the intrinsic and extrinsic camera parameters of the stereo system, and having found corresponding points in the left and right camera images, this chapter proceeds to show how a 3D point cloud can be created via triangulation.

A watertight 3D mesh is the expected input for many CAD or CFD packages where engineers will seek to carry out further analysis tasks. For this reason, this chapter outlines the process of upgrading and cleaning a 3D point cloud to a watertight mesh. This opens the possibility of performing a host of upstream analysis tasks, such as computational fluid dynamics (CFD) simulations using the reconstructed 3D geometries.

REFERENCES

Alagoz, B. Baykant. "Obtaining depth maps from color images by region based stereo matching algorithms." *arXiv preprint arXiv:0812.1340* (2008).

Birchfield, Stan, and Carlo Tomasi. "A pixel dissimilarity measure that is insensitive to image sampling." *IEEE Transactions on Pattern Analysis and Machine Intelligence* 20, no. 4 (1998): 401–406.

Chambers, Lily D., Keith R. Stokes, Frank C. Walsh, and Robert JK Wood. "Modern approaches to marine antifouling coatings." *Surface and Coatings Technology* 201, no. 6 (2006): 3642–3652.

Cocito, S., S. Sgorbini, A. Peirano, and M. Valle. "3-D reconstruction of biological objects using underwater video technique and image processing." *Journal of Experimental Marine Biology and Ecology* 297, no. 1 (2003): 57–70.

Faugeras, Olivier D., Q.-T. Luong, and Stephen J. Maybank. "Camera self-calibration: Theory and experiments." In *European conference on computer vision*, pp. 321–334. Springer, Berlin, Heidelberg, 1992.

Giachetti, Andrea. "Matching techniques to compute image motion." *Image and Vision Computing* 18, no. 3 (2000): 247–260.

Hartley, Richard, and Andrew Zisserman. *Multiple view geometry in computer vision*. Cambridge University Press, 2003.

Kanade, Takeo, and Masatoshi Okutomi. "A stereo matching algorithm with an adaptive window: Theory and experiment." *IEEE Transactions on Pattern Analysis and Machine Intelligence* 16, no. 9 (1994): 920–932.

Kazhdan, Michael, and Hugues Hoppe. "Screened poisson surface reconstruction." *ACM Transactions on Graphics (ToG)* 32, no. 3 (2013): 29.

Lowe, David G. "Object recognition from local scale-invariant features." In *Computer vision, 1999. The proceedings of the seventh IEEE international conference on*, vol. 2 (19991), pp. 1150–1157.

Nguyen, Thanh Minh, and QM Jonathan Wu. "A fuzzy logic model based Markov random field for medical image segmentation." *Evolving Systems* 4, no. 3 (2013): 171–181.

O'Byrne, Michael, Vikram Pakrashi, Franck Schoefs, and Bidisha Ghosh. "A comparison of image based 3D recovery methods for underwater inspections." In *EWSHM-7th European Workshop on Structural Health Monitoring*. 2014.

O'Byrne, Michael, Vikram Pakrashi, Bidisha Ghosh, and Franck Schoefs. "Evaluation of camera calibration techniques for quantifying deterioration." In *Proceedings of the Civil Engineering Research in Ireland Conference (CERI 2016)*. 2016.

Olofsson, Anders. "Modern stereo correspondence algorithms: Investigation and evaluation." (2010). PhD thesis, Linkoping University, Sweden, pp. 5–86, 2010. Retrieved from http://liu.diva-portal.org/smash/get/diva2:328101/FULL TEXT02.pdf.

Papadakis, Nicolas, and Vicent Caselles. "Multi-label depth estimation for graph cuts stereo problems." *Journal of Mathematical Imaging and Vision* 38, no. 1 (2010): 70–82.

Sappa, Angel Domingo, Fadi Dornaika, Daniel Ponsa, David Gerónimo, and Antonio López. "An efficient approach to onboard stereo vision system pose estimation." *IEEE Transactions on Intelligent Transportation Systems* 9, no. 3 (2008): 476–490.

Scharstein, Daniel, and Richard Szeliski. "A taxonomy and evaluation of dense two-frame stereo correspondence algorithms." *International Journal of Computer Vision* 47, no. 1–3 (2002): 7–42.

Sun, Jian, Heung-Yeung Shum, and Nan-Ning Zheng. "Stereo matching using belief propagation." In *European Conference on Computer Vision*, pp. 510–524. Springer, Berlin, Heidelberg, 2002.

Sun, Jian, Hongyan Zhu, Zongben Xu, and Chongzhao Han. "Poisson image fusion based on Markov random field fusion model." *Information Fusion* 14, no. 3 (2013): 241–254.

Sun, Jian, Nan-Ning Zheng, and Heung-Yeung Shum. "Stereo matching using belief propagation." *IEEE Transactions on pattern analysis and machine intelligence* 25, no. 7 (2003): 787–800.

Szeliski, Richard, Ramin Zabih, Daniel Scharstein, Olga Veksler, Vladimir Kolmogorov, Aseem Agarwala, Marshall Tappen, and Carsten Rother. "A comparative study of energy minimization methods for markov random fields." In *European conference on computer vision*, pp. 16–29. Springer, Berlin, Heidelberg, 2006.

Tavera-Vaca, C.-A., D.-L. Almanza-Ojeda, and M.-A. Ibarra-Manzano. "Analysis of the efficiency of the census transform algorithm implemented on FPGA." *Microprocessors and Microsystems* 39, no. 7 (2015): 494–503.

Wang, Kelvin CP, Zhiqiong Hou, and Weiguo Gong. "Automated road sign inventory system based on stereo vision and tracking." *Computer-Aided Civil and Infrastructure Engineering* 25, no. 6 (2010): 468–477.

Xu, Xiao, Renzheng Che, Rui Nian, Bo He, Meimei Chen, and Amaury Lendasse. "Underwater 3D object reconstruction with multiple views in video stream via structure from motion." In *OCEANS 2016-Shanghai*, pp. 1–5. IEEE, 2016.

Zhang, Zhengyou, Quang-Tuan Luong, and Olivier Faugeras. "Motion of an uncalibrated stereo rig: Self-calibration and metric reconstruction." *IEEE Transactions on Robotics and Automation* 12, no. 1 (1996): 103–113.

Zhuang, Yan, and Wei Wang. "Hierarchical adaptive stereo matching algorithm for obstacle detection with dynamic programming." *Journal of Control Theory and Applications* 7, no. 1 (2009): 41–47.

Chapter 8

Repository and interpretation

8.1 INTRODUCTION

The previous chapters introduced a range of image-based techniques for assessing various forms of damage on marine structures. There are many other algorithms that can be used for the same purposes, and it can be expected that the performance of each algorithm will vary under different circumstances. Currently, there exists no standardized approach for choosing a technique that is best-equipped to deal with a given set of circumstances. The major factors that influence technique performance are the onsite visibility conditions, for which turbidity and lighting are the most notable, and also the photometric and geometric properties of the object under inspection as well as the nature of the damage. A key question that confronts inspectors, therefore, is how the techniques at their disposal can be expected to perform under a variety of on-site visibility conditions. To address this question, this chapter presents a repository driven approach to comprehensively map the effect of turbidity and lighting on the performance of image-based techniques. Turbidity and lighting are considered as these are the primary factors that influence the degree of visibility.

The repository is called "Underwater Lighting and Turbidity Image Repository" (ULTIR). It consists of images featuring various damage forms and material types that have been photographed under controlled lighting and turbidity levels, reflective of real-world operating conditions. ULTIR may be accessed through a user-friendly web interface (available at http://www.ultir.net). From here, annotated images of submerged specimens featuring artificial and real damage can be viewed and downloaded. Ground-truth data is also provided that shows the true locations of damage. ULTIR consists of three damage categories relating to (1) 1D crack detection, (2) 2D surface damage detection, and (3) 3D shape recovery using stereo-vision. The imagery contained within each category was captured under three lighting levels and three turbidity levels, resulting in nine images for each specimen. The specimens thus cover a wide range of geometric and photometric properties. The performance of damage assessment methods can be evaluated under realistic operating conditions for each category and compared with the performance of existing methods.

The purpose of ULTIR is to give inspectors and researchers an insight into the relationship between underwater visibility and the performance of image-processing based techniques. An understanding of this relationship has substantial practical benefits. First and foremost, it enables inspectors to assess the viability of adopting image-processing based approaches prior to an inspection and helps them to identify the limits at which image based methods begin to produce unacceptably poor results. Second, the information gleaned from applying algorithms to the imagery in ULTIR can be used during actual inspections to create conditions that are most conducive to good detection such as the provision of suitable lighting—something that is easily attainable. Finally, this repository can help inspectors to choose a suitable algorithm that performs well under the given operating conditions.

This chapter describes the repository and outlines the process of evaluating image-based methods. Several common image methods, along with the techniques described in Chapters 5–7, are applied to imagery in the repository.

8.2 EXPERIMENTAL SET-UP

The imagery contained in ULTIR was generated from experiments conducted in an underwater setting. This section describes the experimental set-up and the nature of the imagery in the repository.

The images in ULTIR were generated from experiments that were conducted in a water basin. These experiments were run in two phases. The set-up for the first phase entailed having a single underwater camera focused on damaged specimens. This phase produced the imagery for the crack and surface damage sections of the repository. The second phase employed on a dual-camera set-up as shown in Figure 8.1. This phase produced the stereo imagery for the 3D shape information section of the repository.

The turbidity levels were regulated using finely sieved kaolin. Varying amounts of kaolin were added to the water tank to produce turbidity levels of 0 NTU (no kaolin), 6 NTU, and 12 NTU. Kaolin is a soft white clay consisting principally of the mineral kaolinite, which goes into suspension when mixed with water. Regular stirring was carried out to ensure kaolin remained in suspension and was uniformly distributed.

The light levels were monitored using a lux meter. Ambient lighting and a set of artificial lights were used to produce light levels of 100, 1000, and 10000 lux. The light level was measured at the same position just above the specimen. While the illumination over the surface of a specimen will vary somewhat as a result of the slightly different distances from the light sources, the lux values span multiple orders of magnitude so slight variations within a given light level are negligible compared to differences between levels.

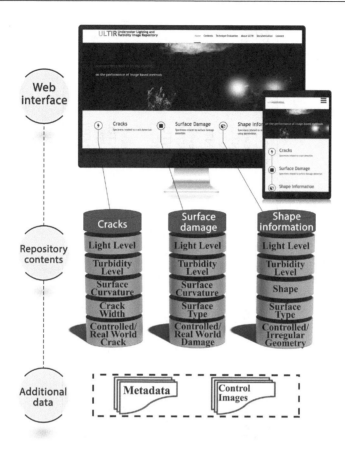

Figure 8.1 Plan view of the experiment set-up.

8.2.1 Contents of the repository

ULTIR contains a large and diverse array of images. This is necessary for comprehensive validation of algorithms and ensures that the repository has a wide applicability. A full breakdown of the contents of the repository is summarized in Table 8.1. The images are characterized by several attributes; along with the turbidity and lighting levels and the type of damage under consideration, the images are characterized by attributes such as the surface type and object curvature. The significance of each attribute is discussed in following sub-sections.

Although there are a similar number of specimens used in each category, it may be noted from Table 8.1 that there are significantly more images in the 3D shape section of the repository than in the crack and surface damage sections. This is due to the fact that recovery of 3D shape requires multiple photographs to be captured from different camera angles. In total,

Table 8.1 Breakdown of the contents of the repository

Category	No. of specimens	Level of control	Surface type	Shape/Curvature	Light levels	Turbidity levels	No. of images
Cracks	9	7 controlled, 2 real cracks	8 concrete, 1 textured concrete	Surface curvature: 4 flat, 5 curved	3 levels: 100 lux, 1000 lux, 10000 lux	3 levels: 0 NTU, 6 NTU, 12 NTU	81
Surface damage	10	9 controlled, 1 real damage	4 concrete, 3 textured concrete, 3 metallic	Surface curvature: 4 flat, 3 cylindrical, 3 spherical	3 levels: 100 lux, 1000 lux, 10000 lux	3 levels: 0 NTU, 6 NTU, 12 NTU	90
3D shapes	12	9 controlled, 3 irregular shapes	4 concrete, 4 metallic, 3 plastic, 1 rubber	3 cubes, 3 cylinders, 3 spheres, 3 irregular shapes	3 levels: 100 lux, 1000 lux, 10000 lux	3 levels: 0 NTU, 6 NTU, 12 NTU	(108 × 8) 864

four stereo pairs (comprising two photographs each from the left and right cameras) were captured from various sides around the specimens by turning the specimens 45° each time to enable full 3D shape recovery, thereby producing eight images per specimen under each visibility condition.

8.2.2 Controlled and partially controlled images

The specimens are classified as either controlled or partially controlled. For the controlled specimens, the "damage" is artificially created. Introducing artificial damage in such a manner is an established way of accurately evaluating the performance of damage detection techniques as it reduces measurement uncertainties that generally arise when assessing real-world instances of damage where uncertainties and vagaries are usually present (Hearn & Testa, 1991). The partially controlled specimens feature real-world instances of damage or are irregularly shaped objects. In this case, precise knowledge of the size of the damage is not known beforehand and must be visually identified by a human observer from the images. The visually segmented images then act as the control. Approximately 80% of the repository is made up of controlled specimens while the remainder consists of partially controlled specimens. The controlled specimens have the primary purpose of algorithm validation, while the partially controlled specimens are intended for testing.

8.2.3 Damage type

The repository is partitioned into three categories according to the nature of the damage. These three categories relate to cracks, surface damage, and 3D shape information. Specimens from each category are presented and described in the following sub-sections.

8.2.3.1 Cracks

Image-based crack detection algorithms work by identifying features of cracks such as their narrow shape and their lower brightness in comparison to the surroundings. For the controlled part of the repository, cracks were simulated by placing two halves of a split concrete specimen by a fixed distance. The fixed distance or the controlled cracks varied in width from 1 mm to 5 mm. These simulated cracks share all the features of real cracks such as the fine structure and lower brightness along the interior of the crack. Each of the cracked specimens in the repository is shown in Figure 8.2. The images shown were captured under medium lighting (1,000 lux) and in clear water (0 Nephelometric Turbidity Units [NTU]) conditions. Full details of illuminance and turbidity values are discussed in Section 8.2.4. The uncontrolled cracks were real cracks introduced by compression testing of concrete specimens. The ground truth images for

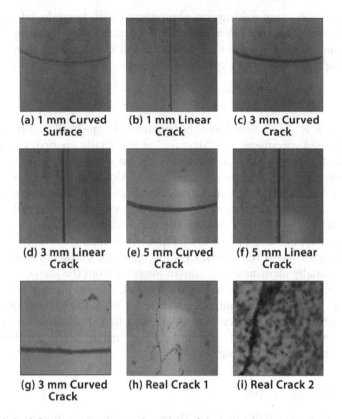

(a) 1 mm Curved Surface   (b) 1 mm Linear Crack   (c) 3 mm Curved Crack

(d) 3 mm Linear Crack   (e) 5 mm Curved Crack   (f) 5 mm Linear Crack

(g) 3 mm Curved Crack   (h) Real Crack 1   (i) Real Crack 2

Figure 8.2 (a–i) Specimens in the crack section of the repository.

the uncontrolled cracks, which are assumed to show the true location and extend of the cracks, had to be identified by a human observer.

8.2.3.2 Surface damage

Marine structures are affected by a wide range of visible damage forms such as corrosion, scour, erosion, leaching, spalling, impact damage, etc. These damages come in an array of shapes and sizes and pose varying levels of significance to the health of a structure. While some damage forms such as pitting corrosion are generally detected using color information, other damages such as spalling and erosion may be more separable based on their textural properties in comparison to the undamaged surface. The contents of the repository reflect this variety by including specimens with different surface textures. For the controlled part of the repository, damage is simu- lated by applying a typically rust-colored emulsion to standard geometric shapes such as squares and rectangles of known dimensions on different surfaces such as metallic and concrete surfaces and on flat, cylindrical,

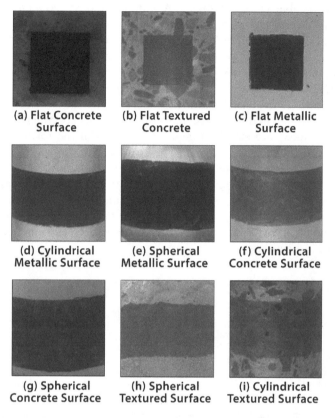

Figure 8.3 (a–i) Specimens in the surface damage section of the repository.

and round surfaces. Specimens in the controlled part of the repository are shown in Figure 8.3. These images were captured under medium light (1,000 lux) and low turbidity (0 NTU) conditions.

8.2.3.3 Shape information

Recovering shape information is a challenging but useful task and has wide applicability in many areas of Structural Health Monitoring (SHM) such as defect detection (Limei et al., 2005) and accurate crack reconstruction (Li & Lowther, 2012) in top-side inspections. The most popular approaches for recovering 3D shape using camera systems are by stereo photography and structure from motion (SfM), as discussed in Chapter 7. Stereo imaging relies on two synchronized cameras that photograph the scene from two slightly different vantage points. The captured images from both cameras are collectively known as a stereo pair. By examining the relative positions of objects in each image, 3D information can be extracted. Stereo-matching algorithms, or stereo-correspondence algorithms, are used to automatically find the same objects in each image. A wide range of stereo matching algorithms have been developed, and choosing the best algorithm is important as the final 3D reconstruction is heavily reliant on correctly matching corresponding points. Additionally, a key stage of the SfM pipeline is the dense matching phase, which uses standard stereo matching algorithms. As such, while ULTIR is principally geared toward stereo imaging, it may also be of value for testing SfM based solutions.

ULTIR is populated by stereo pairs of numerous specimens. The controlled specimens are standard geometric shapes in the form of spheres, cubes, and cylinders, as shown in Figure 8.4. The controlled specimens were chosen as these primitive shapes are the building blocks for more complex shapes. Therefore, it is important to get a fundamental understanding of how stereo-matching algorithms handle the various curvatures. The uncontrolled specimens are irregular shapes with more intricate depth variations, which is more reflective of what would be encountered during real-world inspections.

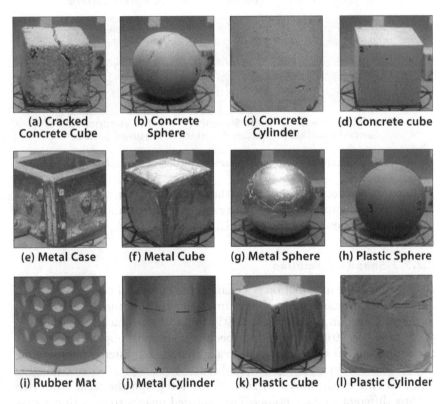

(a) Cracked Concrete Cube   (b) Concrete Sphere   (c) Concrete Cylinder   (d) Concrete cube

(e) Metal Case   (f) Metal Cube   (g) Metal Sphere   (h) Plastic Sphere

(i) Rubber Mat   (j) Metal Cylinder   (k) Plastic Cube   (l) Plastic Cylinder

Figure 8.4   (a–l) Specimens in the surface damage section of the repository. These images were captured under medium light (1,000 lux) and low turbidity (0 NTU) conditions.

8.2.4 Turbidity and lighting

Image quality is assumed to be chiefly affected by luminosity, sharpness (focus accuracy), contrast, and noise. These quality factors are directly related to the on-site operating conditions, for which lighting and turbidity are the most influential (Mahiddine et al., 2012). Underwater imaging is severely hampered by turbidity, which results in reduced contrast, loss of details, and color alteration. Turbidity can be caused by organic particles, such as decomposed plant and animal matter, and algae; or by inorganic particles such as silt and clay. In rivers and lakes, the level of turbidity can fluctuate due to a number of factors such as heavy rains and urban run-off, landslides and bank erosion, algae blooms, interference during the inspection process (e.g., disturbing sediment on the river bed), and human activities such as construction. Given these factors, significant variations in turbidity can be expected. For example, the measured turbidity in the lower Waitaki River, New Zealand, ranges from 1.2 NTU (Nephelometric Turbidity Units) to 23 NTU (Graham, 1990). With this in mind, inspectors may wish to schedule inspections for periods when these factors are expected to cause minimum increases in turbidity levels. In the open ocean, turbidity is affected mostly by seasonal phytoplankton blooms; however, it is generally low.

For ULTIR, three levels of turbidity were chosen: 0 NTU, 6 NTU, and 12 NTU. To put these values in context, clear water has a turbidity of 0 NTU, water that is visibly cloudy has a turbidity of 6 NTU, while water that is murky has a turbidity of 25 NTU. A cut-off point of 12 NTU was chosen for ULTIR as it becomes increasingly difficult to interpret and extract useful information from images beyond this point, and as a consequence, image-based methods quickly lose validity as a quantitative inspection tool. Additionally, the turbidity of many rivers and water bodies lies within the 0–12 NTU range. As an example, Grand River in Michigan is reported to have a turbidity range of 2–9 NTU, while Spring Lake in New Jersey has a turbidity range of 0.1–4 NTU (GVSU, 2013). The lower limit of 0 NTU was chosen as this represents the best-case scenario and it shows the level of performance can be achieved in the absence of any turbidity, while 6 NTU was a natural choice as the intermediate turbidity level as it is the midpoint of the 0–12 NTU range.

Turbidity can be measured on site using a digital nephelometer or by using a Secchi disk, which is a black and white disk lowered into the water until it is no longer visible. The depth of the disk is a measure of the transparency of the water, which is inversely related to the turbidity. This has the advantage of being a quick, inexpensive, and simple approach for measuring turbidity. Additionally, many water bodies, such as the River Lee in Ireland, are increasingly being monitored by in-situ sensor networks, which provide near real-time data on water quality parameters including turbidity (Lawlor et al., 2012).

While lowering the on-site turbidity levels is generally not possible, the effects of high turbidity on image quality can be partially offset by moving the camera closer to the subject. In high-turbid/murky waters, for example, above 100 NTU, an object may disappear if it is separated by just a few centimeters from the observer, while in clearer waters, such as in the open ocean where the turbidity is often close to 0 NTU, an object can be tens of meters away from an observer and still remain visible. The imagery in ULTIR is generated with the cameras and specimens kept at a fixed distance of 80 cm apart. This distance is regarded to be a practical distance that underwater divers would find to be a reasonable compromise between capturing enough of the scene without overly sacrificing detail and clarity.

Lighting also plays a pivotal role in achieving good visibility. Ambient light may be sufficient for near-surface inspections; however, it is unlikely to be sufficient at greater depths at which point artificial light will become necessary. Three light levels were used: 100 lux, 1000 lux, 10000 lux. To put this in perspective, the approximate level of light, or illuminance, on a very dark overcast day is 100 lux, a moderately overcast day is 1000 lux, and full daylight (not direct sunlight) is 10,000–25,000 lux (Schlyter, 2009). The lux is the SI unit of illuminance that measures the intensity of light that strikes a surface, as perceived by the human eye. The illuminance can be measured using a lux meter. A specimen from each section of the repository is shown under varying lighting and turbidity levels in Figures 8.5–8.7.

It may be observed from Figures 8.5–8.7 that the featured specimens are easily distinguishable under some conditions (e.g., medium light with low turbidity), while in other cases, the specimens appear clouded (e.g., low light with high turbidity). While increasing turbidity has a worsening effect on image quality, the relationship between light intensity and image quality is more complicated. A dark, poorly lit scene is clearly unfavorable for the task of visual interpretation of damage; however, too much lighting can also create problems. With reference to the metallic specimen shown in Figure 8.6, it can be seen that lighting complexities, such as bright spots, become apparent at the highest lighting level. This obscures detail in affected regions of the image. Furthermore, high lighting in combination with high turbidity produces a lot of backscatter resulting in a hazy-glow effect, which also masks detail. This phenomenon is evident in Figure 8.6(i). The interplay between lighting and turbidity means that these must be considered jointly for the purpose of assessing the performance of image-based algorithms.

8.2.5 Surface type

The main construction materials for marine structures are concrete and metal. These materials have different photometric properties. Metallic surfaces often appear shiny while concrete surfaces generally appear to be

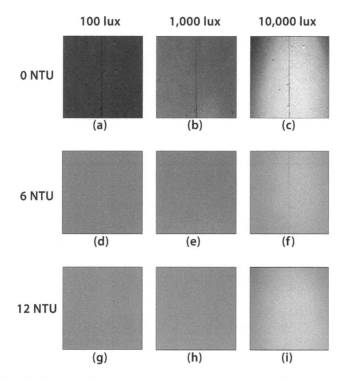

Figure 8.5 (a–i) Controlled crack specimen shown under varying lighting and turbidity conditions. Columns: Low (100 lux), Medium (1000 lux), High (10000 lux) Light. Rows: Low (0 NTU), Medium (6 NTU), High (12 NTU) Turbidity.

dull. High specular reflections pose problems for image analysis algorithms as the shine masks details in the scene and create artifacts that could mislead algorithms. This is especially problematic when strong artificial light sources are used.

The texture of a surface is another important property that influences image analysis. In the case of stereo matching, for example, finding matching points on a smooth/uniform surface is more ambiguous than matching points on a richly textured surface. Additionally, the surface texture may be a consideration when deciding on what type of 2D damage detection algorithm to use. If damage is more differentiable from the background based on texture than by color, then it may be worthwhile segmenting based on texture. Finally, crack detection algorithms applied to surfaces with a rippled texture can produce a lot of false alarms (e.g., the specimen shown in Figure 8.2(i)). To cater for these issues, ULTIR contains metallic surfaces with high specular reflectivity as well as diffuse concrete surfaces with various textural finishes.

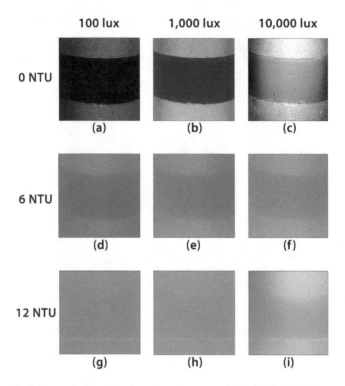

Figure 8.6 (a–i) Controlled surface damage specimens under varying lighting and turbidity conditions. Columns: Low (100 lux), Medium (1000 lux), High (10000 lux) Light. Rows: Low (0 NTU), Medium (6 NTU), High (12 NTU) Turbidity.

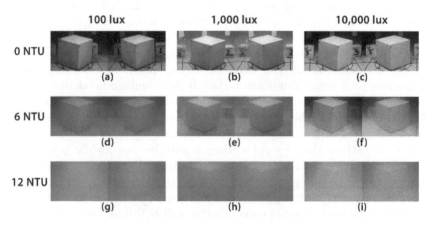

Figure 8.7 (a–i) Stereo pairs featuring a concrete cube shown under varying lighting and turbidity levels. Columns: Low (100 lux), Medium (1000 lux), High (10000 lux) Light. Rows: Low (0 NTU), Medium (6 NTU), High (12 NTU) Turbidity.

8.3 ONLINE PORTAL OF ULTIR

Users of the repository will first encounter the web interface, which is depicted in Figure 8.8.

The design of the web interface is focused on facilitating navigation through the repository. An extensive overview of the repository and a user guide is available at the ULTIR website. Some of the key web pages are shown in Figure 8.9.

The documentation and user guide, shown in Figure 8.9(a), explains the naming convention and advises on how to navigate through the repository. The naming convention was adopted to succinctly convey important information such as the specimen description, turbidity level, and lighting level. Figure 8.9(b) presents an overview of all images in each category of the repository, in this case, the crack repository, and reveals details about the nature of the crack such as the crack width. Clicking on any thumbnail in this list will bring up the full assortment of associated images for each turbidity and lighting level, along with a binary control image that shows the damaged region in white and the undamaged in black, as shown in Figure 8.9(c). Metadata such as the aperture, focal length, and ISO remain embedded in the images. The labelling schema encodes concise contextual information about the specimen under consideration such as the size and shape properties, the material properties, and a description of the damage.

The technique evaluation page is shown in Figure 8.9(d). The page allows inspectors to identify and adopt techniques that work well for a given situation, as well as allowing algorithm developers to compare their techniques against existing state-of-the-art methods.

Figure 8.8 Layout and contents of the repository.

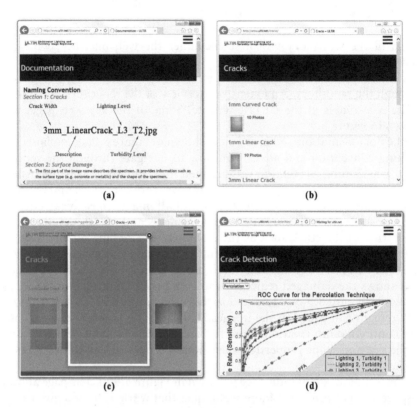

Figure 8.9   (a) ULTIR documentation and user guide, (b) crack page, which lists the specimens and nature of the crack, (c) sample image from the crack part of the
repository captured in low light and high turbidity, and (d) technique evaluation page.

8.4 ROC-BASED PERFORMANCE
   EVALUATION OF ALGORITHMS

This section compares some of the main algorithmic approaches related to
each category of ULTIR, namely: crack detection, surface damage detection, and 3D shape recovery. The purpose of this section is to characterize
the performance of image algorithms under a given set of environmental
conditions and to identify state-of-the-art methods. A number of algorithms
are applied to the datasets depicted in Figure 8.5–8.7. The performance
of submitted techniques is evaluated and ranked using receiver operating
characteristic (ROC) curves.

The ROC curves offer a convenient way of characterizing the performance
of NDT methods under various environmental conditions (Rouhan & Schoefs,
2003) and have been expanded to image detection (Pakrashi et al., 2010).
A ROC curve is a plot of the true positive rate (sensitivity) versus the false

positive rate (1-specificity) that are obtained when varying an input parameter for a given technique. Sensitivity, or the probability of detection in the field of probability space and decision theory, measures the proportion of pixels that are correctly identified as representing damage. Specificity measures the proportion of non-damaged pixels that are correctly identified as representing non-damage. Throughout this book, the sensitivity and specificity are determined by comparing the damaged regions detected using an image-based technique with a visually segmented image. The visually segmented image is created by a human operator who must manually identify damaged regions in an image. This visually segmented image acts as the control as it is assumed it shows the true extent of damage. The visually segmented image only needs to be created when it is wished to gauge the performance levels of the technique under scrutiny. Each (1-specificity, sensitivity) pair forms a coordinate in the ROC space that corresponds to a particular decision threshold.

The following example demonstrates the use of ROC curves to characterize the performance of a simple thresholding technique for detecting a rust stain. Here, the thresholding technique is applied several times to an input image (Figure 8.10(a)) using various thresholds. For each threshold, the detected damaged region (represented by white pixels) is compared against the control image (as shown in Figure 8.10(b)) and the (1-specificity, sensitivity) pair is computed.

The α-δ method is employed to find the optimum threshold value that maximizes detection. This method provides a measure of how well a parameter can distinguish between two diagnostic groups (i.e., damaged region/non-damaged region) (Baroth et al., 2011; Schoefs et al., 2012). It relies on calculating the angle, α, and the Euclidean distance, δ, between the best performance point, defined as an ideal NDT technique with 100% sensitivity and 100% specificity and the considered point to give a measure of the performance of the considered point. As this book does not deal with risk analysis where the shape of the ROC acts as a key factor, only the delta, δ, parameter is required as a measure of performance. A low value for δ

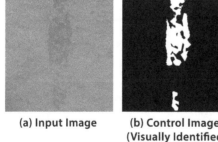
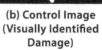

| (a) Input Image | (b) Control Image (Visually Identified Damage) | (c) Detected Damage at the Optimum Threshold Value (0.6) |

Figure 8.10 (a) Input image, (b) control image, and (c) detected damage at the optimum threshold value.

is indicative of a good performance. Therefore, the closer the ROC curve is to the upper left corner of the plot, the higher the overall accuracy of the technique. The minimum value of δ is considered to correspond to the parameter that produces the best detection results. The result of applying the thresholding technique to the input image using the optimum parameter is shown in Figure 8.10(c).

The following code shows how to compute the (1-specificity, sensitivity) pair for each threshold value, and subsequently generate a ROC curve (Figure 8.11).

```
1 % Display Color, Grayscale and Binary Image
2 InputImage = imread('RustStain_L2_T1.tif');
3 % Read in control image and cast as a binary image
4 Control = im2bw(imread('Control.tif'));
5
6 % Apply a simple thresholding method several times, each
  time
```

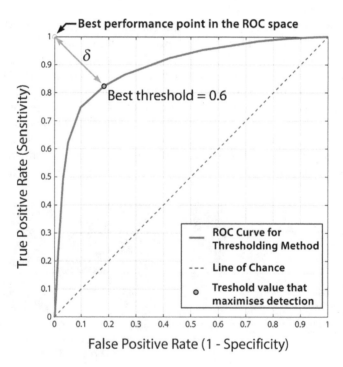

Figure 8.11 ROC curve for a simple thresholding technique.

```
7 % varying the threshold parameter from 0 to 1 in steps
  of 0.1
8 counter = 1;
9 for level = 0:0.01:1
10 BW = 1 - im2bw(InputImage, level);%Perform thresholding
11 % TP(level) = num. of samples correctly classified as
   positive
12 TP(counter) = sum((BW(:) == 1).*(Control(:) == 1));
13 % TN(level) = num. of samples correctly classified as
   negative
14 TN(counter) = sum((BW(:) == 0).*(Control(:) == 0));
15 Tvalues(counter) = level; % Keep track of threshold
   values
16 counter = counter + 1;
17 end
18
19 % Defining P = no. of positive samples
20 P = sum(Control(:));% Counts all white pixels in
   control image
21 % N = num. of negative samples (total no. of pixels
   minus P)
22 N = size(Control,1)*size(Control,2) - P;
23 % Express as True Positive Rate and True Negative Rate
24 TPR = TP./P;
25 TNR = TN./N;
26
27 figure, hold on
28 plot(1-TNR,TPR)% Plot ROC curve
29 plot([0,1],[0,1],'.-') % Line of Chance
30
31 % Find the best performance point
32 delta = sqrt((1 - TPR).^2 + (1-TNR).^2); % Euclidean
   distance
33
34 % Highlight this point in the plot
35 ind = find(delta == min(delta),1);
36 quiver(0,1,(1-TNR(ind)),(TPR(ind)-1))
37 plot(1-TNR(ind),TPR(ind),'o')
38 text(1-TNR(ind),TPR(ind), ['Best threshold = ',
39 num2str(Tvalues(ind))])
40 axis square; hold off
41
42 % Display images side by side
43 figure, subplot(1,3,1),imshow(InputImage),title('Input
   Image')
44 subplot(1,3,2), imshow(Control), title('Control Image')
45 subplot(1,3,3), imshow(1 - im2bw(InputImage, 0.6)),
46 title('Detected Damage at Optimum Threshold Value
   (0.6)')
```

The more area under a ROC curve the better is the image-based technique. The straight line that connects the point (0,0) with the point (1,1) is referred to as the line of chance and corresponds to a random testing. Overall, ROC curves provide an intuitive way of characterizing the performance of techniques, as well as providing a way to graphically compare several different techniques applied to the same imagery.

8.4.1 Crack detection

A number of techniques have been devised that are capable of identifying crack-like features that are characterized by their narrow shape and lower brightness in comparison to the surroundings. These include a percolation-based method described in Chapter 5, eigenvalue analysis of the Hessian (Frangi et al., 1998), Kirsch templates (Kirsch, 1971), neural networks (Choudhary & Dey, 2012) and statistical filters (Sinha & Fieguth, 2006). This section applies the first three of these methods to the 1 mm controlled crack data set (as shown in Figure 8.5) from the repository to investigate the effects of changing turbidity and lighting levels on the detection accuracy for each technique.

In practice, the input for these crack detection methods is an image, or a batch of images, that feature visible cracks. The output is a binary image (black and white image) in which the white pixels denote pixels that likely correspond to cracks, while black pixels correspond to the background. Crack detection methods employ various approaches to identify and isolate cracks, some of which are described below. The percolation method is based on tracing out dark pixels within a fixed-sized window, or sub-image, starting at the center point of the window, and spreading out until the boundary of the window is reached. The resulting pattern of dark pixels is analyzed. Cases where narrow or linear patterns are traced out are indicative of cracks, while irregular or radial diffusion patterns typically correspond to the non-cracked background. Percolation methods tend to be robust to noise in the image; however, they are also computationally expensive since they operate on a pixel-by-pixel basis. The eigenvalue analysis of the Hessian method detects narrow crack-like paths by calculating the direction of smallest curvature where there is a minimum change in intensity, which is usually along the crack path. A feature of this method is that it performs simultaneous noise and background suppression when analyzing the local structure in the image. A drawback of the method is that some artifacts are introduced in regions where background fluctuations have line patterns. The Kirsch templates method detects line-like objects using spatial filtering involving templates orientated in eight different directions followed by thresholding. While this method is computationally efficient, the lack of any sophisticated classification criteria means the detection accuracy will most likely suffer. Additionally, spatial filtering will produce a subdued response in cases where a crack is not closely aligned with any of the template orientations.

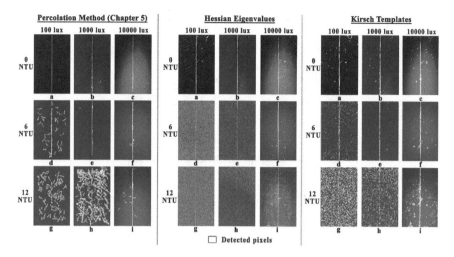

Figure 8.12 Detected cracks corresponding to the images in Figure 8.5.

The results for each of these techniques are shown in Figure 8.12; the performance levels are summarized in Table 8.2.

The ROC curves are generated by applying the techniques to all the cracked specimens in the repository multiple times, each time adjusting the considered technique's critical parameters, and comparing the resulting detected cracks with the corresponding ground truth data in order to obtain a set of sensitivity and specificity pairs.

The ROC curves for all of the specimens under a certain lighting and turbidity condition are then averaged to produce a curve representative of the overall performance of a particular technique at that lighting and turbidity condition. This results in 9 curves for each technique, as shown in Figure 8.13.

It may be observed from the detected cracks in Figure 8.12 that each technique performs quite well for images that feature clear and sufficiently lit scenes. Unsurprisingly, the performance deteriorates when the turbidity levels increase. In the worst visibility conditions—low light and high turbidity—all of the techniques produce poor results suggesting the adoption of image-based crack detection approaches under these conditions is not practical. However, the results show that having high lighting can mitigate the effects of high turbidity. In these situations, the increased absorption and diffusion in turbid water limits the formation of a bright spot that would otherwise impair detection.

Analysis of the δ parameter in Table 8.2 reveals that the percolation-based method performs the best for the medium turbidity level. The cracks are well delineated and there are relatively few misclassified pixels. The eigenvalue analysis of the Hessian method produces a high misclassification rate at the lower light levels, while the Kirsch templates method also produces a lot of small spurious regions. Overall, there is relatively little difference in

Table 8.2 Performance of the crack detection techniques

Image	Condition	Percolation			Hessian eigenvalues			Kirsch templates		
		(DR)	(MCR)	δ	(DR)	(MCR)	δ	(DR)	(MCR)	δ
a	Lighting 1, Turbidity 1	94.9%	0.7%	0.05	96.7%	2.3%	0.04	92.3%	5.3%	0.09
b	Lighting 2, Turbidity 1	91.5%	0.8%	0.09	91.6%	1.2%	0.09	94.1%	4.0%	0.07
c	Lighting 3, Turbidity 1	81.7%	0.7%	0.18	90.7%	3.9%	0.10	96.8%	5.3%	0.06
d	Lighting 1, Turbidity 2	75.0%	3.9%	0.25	58.6%	50.2%	0.65	67.9%	13.2%	0.35
e	Lighting 2, Turbidity 2	88.0%	1.0%	0.12	54.6%	34.1%	0.57	79.4%	4.5%	0.21
f	Lighting 3, Turbidity 2	94.1%	1.2%	0.06	95.4%	7.7%	0.09	92.7%	4.8%	0.09
g	Lighting 1, Turbidity 3	21.5%	8.4%	0.79	44.6%	48.7%	0.74	56.4%	54.2%	0.70
h	Lighting 2, Turbidity 3	54.8%	21.7%	0.50	25.2%	29.8%	0.81	55.3%	36.1%	0.57
i	Lighting 3, Turbidity 3	91.7%	1.7%	0.08	83.7%	24.6%	0.30	92.5%	11.7%	0.14

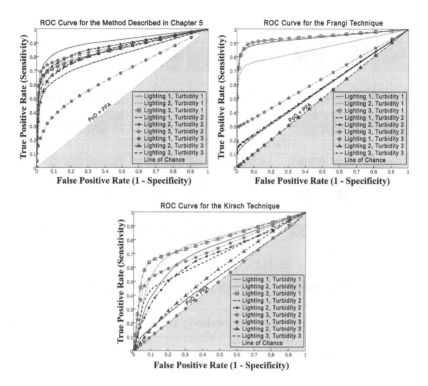

Figure 8.13 Evaluation of crack detection techniques through the use of ROC curves.

terms of detection accuracy for the techniques considered. Instead, the visibility conditions have by far the greatest influence on the output.

8.4.2 Surface damage

Most image-processing based damage detection algorithms consist of segmentation followed by subsequent classification of the segmented regions. Ideally, the segmentation methodology should identify and accurately define all regions of interest in an image while minimizing the inclusion of extraneous regions. In reality, perfect segmentation is difficult to achieve given the inherent chromatic and luminous complexities encountered in natural scenes. Image-processing based techniques include color intensity based methods and texture analysis based methods. Naturally, the techniques in each group are suited to different applications. The effectiveness of color based segmentation algorithms and texture based segmentation algorithms will vary according to the surface and damage type under consideration as certain damages are more separable from the undamaged surface based on either their color or texture attributes. This section assesses the performance of two methods described in Chapter 6: the color based

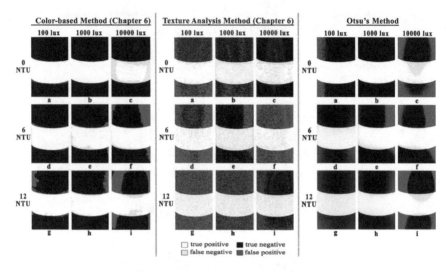

Figure 8.14 Detected damage corresponding to the images in Figure 8.6.

method and the texture analysis method, along with the widely popular Otsu's thresholding method (Otsu, 1979).

These techniques are applied to the imagery in Figure 8.6 and the results are shown in Figure 8.14. The performance levels are quantified in Table 8.3. ROC curves for each technique under the varying lighting and turbidity conditions are shown in Figure 8.15.

These ROC curves are generated by applying the techniques to all of the specimens in the surface damage section of ULTIR multiple times, each time adjusting the considered technique's critical parameters, and comparing the resulting detected damaged region with the corresponding ground truth data in order to obtain a set of sensitivity and specificity pairs. The ROC curves for all of the specimens under a certain lighting and turbidity condition are then averaged to produce a curve representative of the overall performance of a particular technique at that lighting and turbidity condition. This results in 9 curves for each technique.

It may be noted from these results that the color-based method was quite robust—it displayed less sensitivity to the input conditions as evidenced by the close bunching of ROC curves across all lighting and turbidity levels in Figure 8.15. Moreover, it proved effective at locating the presence of damage as well as accurately defining the shape and size of damaged regions. The success of Otsu's method may be explained by the fact that the damaged region is characterized by one predominant color, which is sufficiently distinct from the background. Both Otsu's method and texture analysis had many small spurious regions, unlike the color based method that produced a "cleaner" and more homogenous detection.

Table 8.3 Performance of the damage detection techniques

Image	Condition	Color method			Texture analysis			Otsu's method		
		(DR)	(MCR)	δ	(DR)	(MCR)	δ	(DR)	(MCR)	δ
a	Low Light, Low Turbidity	99.5%	2.1%	0.02	97.2%	49.3%	0.49	98.1%	14.8%	0.15
b	Medium Light, Low Turbidity	97.5%	1.2%	0.03	89.9%	33.1%	0.35	97.5%	1.2%	0.03
c	High Light, Low Turbidity	60.3%	14.2%	0.42	75.5%	15.8%	0.29	76.5%	39.8%	0.46
d	Low Light, Medium Turbidity	98.1%	2.5%	0.03	98.0%	51.3%	0.51	95.7%	22.2%	0.23
e	Medium Light, Medium Turbidity	91.9%	5.7%	0.10	82.6%	20.6%	0.27	92.6%	18.9%	0.20
f	High Light, Medium Turbidity	93.3%	16.7%	0.18	98.8%	61.5%	0.61	94.1%	24.6%	0.25
g	Low Light, High Turbidity	99.2%	7.0%	0.07	90.1%	61.8%	0.63	83.5%	37.5%	0.41
h	Medium Light, High Turbidity	93.4%	6.9%	0.10	83.8%	58.0%	0.60	93.9%	20.9%	0.22
i	High Light, High Turbidity	85.4%	24.0%	0.28	76.8%	34.0%	0.41	82.9%	30.7%	0.35

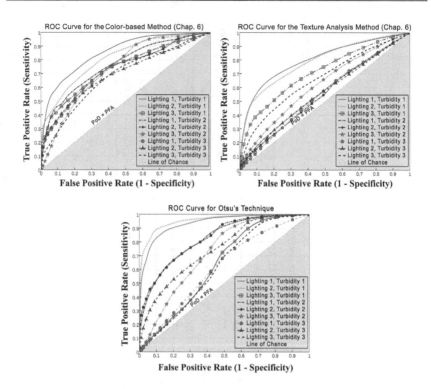

Figure 8.15 Evaluation of damage detection techniques through the use of ROC curves.

The texture based method was effective at locating the presence of damage as may be observed from Figure 8.14; however, it did not perform as well as the color based methods at defining the extent of damage that resulted in high δ values in Table 8.3. The exception to this was for the low turbidity and high light image (Figure 8.6(c)) where the texture analysis outperformed the other methods. While the shiny metallic surface created luminous complexities that misled the color based methods, the bright light "revealed" some of the textural properties of the surface that benefitted the texture analysis technique.

While the visibility conditions had a major impact on the detection of cracks, the performance of surface damage techniques does not rapidly decline with deteriorating visibility condition. Instead, the results indicate that choice of technique is more critical.

8.4.3 3D shape recovery using stereo vision

The correspondence problem is often difficult because of ambiguous correspondences that can lead to false matches. This is especially problematic for uniform surface types whereby the lack of distinct features causes a high number of vague matches. Such scenarios are particularly likely to arise in underwater settings due to the poor visibility conditions.

This section compares the performance of three distinct types of stereo correspondence algorithms. The first is the belief propagation (BP) Markov random field (MRF) method described in Chapter 7. The second method is based on the well-known Birchfield and Tomasi (1998) matching cost that is robust to image sampling. This method compares each pixel in the reference image against a linearly interpolated function of the other image. It does not rely on any smoothness constraints, but rather, the disparity is computed by selecting the minimal (winning) aggregated value at each pixel.

The final method is a region based stereo matching algorithm (Alagoz, 2008), which operates by semi-global error energy minimization by smoothing functions. This method chooses a root point in a region and then grows that point while the energy function remains equal or less than a certain value. Otherwise, a new root point is selected and a separate region starts growing from there. This algorithm employs the sum of absolute differences (SAD) as a similarity measure.

These techniques are applied to the stereo imagery featuring a concrete cube in Figure 8.7 and the resulting disparity maps, or depth maps, are shown in Figure 8.16. A low normalized root-mean-square error (NRMSE) score is indicative of a strong performing stereo correspondence technique as it corresponds to cases where there is a small deviation between the computed disparity map and the ground truth disparity map.

The aggregated NRMSE scores for all of the specimens in the repository under each lighting and turbidity are illustrated in the bar chart in Figure 8.17. A description of the specimens is provided in Table 8.4.

Results show a marked variation across the range of specimens in ULTIR. Specimens, such as the rough concrete cube (specimen no. 1) and the aged metallic case (specimen no. 8), facilitate more reliable stereo matching due to the richly textured nature of these surfaces. This is in contrast with specimens that have more uniform and homogeneous surfaces, such as the plastic cylinder (specimen no. 15), whereby the lack of unique features is seen to inhibit the performance of the stereo algorithms.

It may be observed from the disparity maps in Figure 8.16 that the method described in Chapter 7 produces the best results in most cases. Given that a key feature of this technique is that it explicitly accounts for smoothness

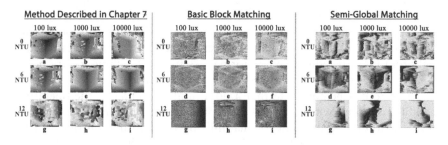

Figure 8.16 **Disparity maps for the stereo imagery in Figure 8.7.**

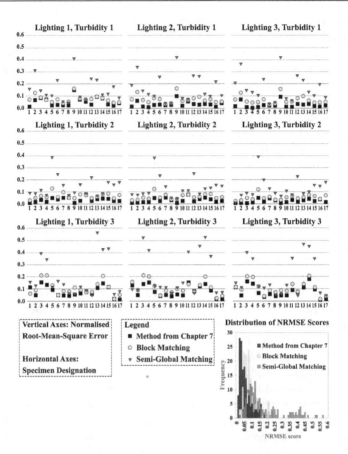

Figure 8.17 The aggregated NRMSE scores for all of the specimens in the repository under each lighting and turbidity.

between neighboring pixels, a reasonable interpretation would be that algorithms also incorporating some form of smoothness term would fare better when applied to images in the ULTIR repository and, ultimately, when performed on images obtained from real-world underwater inspection. The usefulness of a smoothness term appears to be even more pronounced when operating at higher turbidity and light levels at which point some of the surface details become washed out. However, in the worst conditions (i.e., low light and high turbidity), the performances from all techniques are poor. This indicates that the operating limit has been reached.

These results demonstrate that a good understanding of the operating conditions, along with a careful and considered choice of image processing algorithm, is important for maximizing the information that can be obtained from image analysis. This analysis could only be achieved with a resource such as ULTIR.

Table 8.4 Description of specimens

1	Rough Concrete Cube Planar	2	Rough Concrete Cube Diagonal	3	Smooth Concrete Cube Planar	4	Smooth Concrete Cube Diagonal	5	Concrete Cylinder	6	Concrete Sphere
7	Metal Case Diagonal	8	Metal Case Planar	9	Metal Cube Planar	10	Metal Cube Diagonal	11	Metal Cylinder	12	Metal Sphere
13	Plastic Cube Planar	14	Plastic Cube Diagonal	15	Plastic Cylinder	16	Plastic Sphere	17	Rubber lattice		

8.5 CONCLUSIONS

Image-based methods undoubtedly have value as a quantitative inspection tool; however, the performance of these methods is often hampered by the poor visibility and complex underwater light field. A key question for inspectors is how the techniques at their disposal perform under various operating conditions, for which the on-site lighting and turbidity levels are considered the most influential factors. While increasing turbidity adversely impacts technique accuracy, the relationship between the illumination and the technique accuracy is more complicated. A dark, poorly lit scene is clearly unfavorable for the task of detecting damage; however, too much lighting also creates problems, such as bright spots on the surface of objects, which obscures detail. This chapter describes an open-source Underwater Lighting and Turbidity Image Repository (ULTIR). ULTIR is populated with images featuring various damage forms and material types that have been photographed under controlled lighting and turbidity levels and that are reflective of real-world operating conditions.

The aim of ULTIR is to lessen the uncertainty introduced by the onsite visibility conditions, thereby enabling inspectors to make informed decisions when using image-processing methods. The resource also assists researchers when developing and evaluating new or existing image-processing algorithms intended for underwater application. This large and well-annotated dataset paves the way for researchers to develop data-driven algorithms, such as deep learning techniques, that may be more robust to the underwater conditions. Such algorithms have already attracted significant interest in other fields owing to the high performances that they can achieve; however, they have not had an impact in the domain of underwater damage detection as of yet, largely due to the lack of available training data.

This chapter outlines a procedure for evaluating the performance of image-based methods. Several common image methods, as well as the techniques that have been previously described in this book, are applied to images in the repository. The results indicate that the choice of algorithm and the environmental conditions are important factors that affect the overall success of image-based methods, and therefore, these should be given careful consideration when establishing the domain of operation and efficiency of different methods and sites.

REFERENCES

Alagoz, B. Baykant. "Obtaining depth maps from color images by region based stereo matching algorithms." B. *OncuBilim Algorithm and Systems Labs* 8, no. 4 (2008): 1–12.

Baroth, J., Breysse, D., and F. Schoefs. *Construction Reliability*, Wiley, Hoboken, NJ, 2011.

Birchfield, Stan, and Carlo Tomasi. "A pixel dissimilarity measure that is insensitive to image sampling." *IEEE Transactions on Pattern Analysis and Machine Intelligence* 20, no. 4 (1998): 401–406.

Choudhary, Gajanan K., and Sayan Dey. "Crack detection in concrete surfaces using image processing, fuzzy logic, and neural networks." In *Advanced Computational Intelligence (ICACI), 2012 IEEE Fifth International Conference on*, pp. 404–411. IEEE, 2012.

Frangi, Alejandro F., Wiro J. Niessen, Koen L. Vincken, and Max A. Viergever. "Multiscale vessel enhancement filtering." In *International Conference on Medical Image Computing and Computer-Assisted Intervention*, pp. 130–137. Springer, Berlin, Heidelberg, 1998.

Graham, A. A. "Siltation of stone-surface periphyton in rivers by clay-sized particles from low concentrations in suspension." *Hydrobiologia* 199, no. 2 (1990): 107–115.

Hearn, George, and Rene B. Testa. "Modal analysis for damage detection in structures." *Journal of Structural Engineering* 117, no. 10 (1991): 3042–3063.

Kirsch, Russell A. "Computer determination of the constituent structure of biological images." *Computers and Biomedical Research* 4, no. 3 (1971): 315–328.

Lawlor, Antóin, Javier Torres, Brendan O'Flynn, John Wallace, and Fiona Regan. "DEPLOY: A long term deployment of a water quality sensor monitoring system." *Sensor Review* 32, no. 1 (2012): 29–38.

Li, Min, and David Lowther. "NDT sensor design optimization for accurate crack reconstruction." *COMPEL-The International Journal for Computation and Mathematics in Electrical and Electronic Engineering* 31, no. 3 (2012): 792–802.

Limei, Song, Zhou Xinglin, and Xu Kexin. "Three-dimensional defect detection based on single measurement image." *Acta Optica Sinica* 25, no. 9 (2005): 1195.

Mahiddine, Amine, Julien Seinturier, Daniela Peloso Jean-Marc Boi, Pierre Drap, Djamel Merad, and Luc Long. "Underwater image preprocessing for automated photogrammetry in high turbidity water: An application on the Arles-Rhone XIII roman wreck in the Rhodano river, France." In *Virtual Systems and Multimedia (VSMM), 2012 18th International Conference on*, pp. 189–194. IEEE, 2012.

Otsu, Nobuyuki. "A threshold selection method from gray-level histograms." *IEEE Transactions on Systems, Man, and Cybernetics* 9, no. 1 (1979): 62–66.

Pakrashi, Vikram, Franck Schoefs, Jean Bernard Memet, and Alan O'Connor. "ROC dependent event isolation method for image processing based assessment of corroded harbour structures." *Structures & Infrastructure Engineering* 6, no. 3 (2010): 365–378.

Rouhan, Antoine, and Franck Schoefs. "Probabilistic modeling of inspection results for offshore structures." *Structural Safety* 25, no. 4 (2003): 379–399.

Schlyter, Paul. "Radiometry and photometry in astronomy." *Available: Stjarnhimlen. se/comp/radfaq.html* (2009).

Schoefs, F., Boéro, J., Clément, A., and B. Capra. "The αδ method for modelling expert judgment and combination of NDT tools in RBI context: Application to marine structures." *Structure and Infrastructure Engineering: Maintenance, Management, Life-Cycle Design and Performance (NSIE). Monitoring, Modelling and Assessment of Structural Deterioration in Marine Environments* 8(Special Issue), (2012): 531–543.

Sinha, Sunil K., and Paul W. Fieguth. "Automated detection of cracks in buried concrete pipe images." *Automation in Construction* 15, no. 1 (2006): 58–72.

Chapter 9

Examples of future applications

9.1 INTRODUCTION

This chapter explores technologies that are promising in terms of shaping the future of underwater inspections for built infrastructure. Improvements that can be made to current methodologies are discussed and new research directions are outlined. For inspectors and engineers to get the full benefit out of image-processing techniques, it is important that they can easily incorporate the results from image-processing into their overall assessment. In this vein, this chapter presents a workflow for feeding 3D data, obtained through stereo imaging, into a computational fluid dynamics (CFD) software package for further analysis.

Virtual reality (VR) is becoming increasingly popular. This chapter examines the value that VR can bring to underwater inspections and describes how immersive virtual underwater scenes can assist divers and inspectors when planning and conducting real-world inspections. The role that spherical cameras can play in underwater inspections is investigated. Spherical cameras record the whole environment around the camera, which reduces the dependency on divers to notice and purposefully photograph instances of damage. The advantages and potential problems associated with this form of imaging are discussed.

This chapter also looks at algorithmic advances in the field of image-processing. Specifically, deep learning techniques are discussed as these techniques continue to attract significant attention. In the past, deep learning techniques were largely restricted due to a lack of training data and limited computational power. While computational power has drastically improved over the years, there is still a scarcity of reliable training data for many underwater damage assessment applications. Ways to generate more training data from existing damage datasets in the form of photographs and videos obtained from built infrastructure inspection are suggested. This chapter also touches on video analysis as an extension to standard image analysis methods. Video is more than just a set of individual frames and contains temporal information that is useful for a wide array of applications in engineering and Structural Health Monitoring (SHM).

As an example application, video tracking is employed to characterize the dynamic response of a river-spanning pedestrian bridge. On top of this, this chapter looks at how smartphones can act as viable video acquisition devices, thanks in a large part to the ever-improving image quality of their camera sensors.

The final part of this chapter deals with how archived inspection imagery can be used by inspectors/divers for both current assessments and to guide future inspections. Ways in which archived inspection imagery can be quantitatively analyzed despite missing some metadata or auxiliary information are highlighted.

9.2 INTEGRATING IMAGE DATA INTO SUBSEQUENT ANALYSES

The quantitative nature of the data obtained from image-processing naturally lends itself to many high-level analysis tasks. It is important that engineers can fully utilize the output from image-processing techniques and seamlessly integrate it into more specialized software packages for further analysis. To demonstrate this process, this section describes a workflow for reconstructing 3D models of underwater structural components using stereo imaging, and then, feeding this 3D data into a computation fluid dynamics (CFD) suite where global loading assessment can be carried out.

It is quite common for 3D CFD simulation models to be developed so as to predict the wave loading on offshore structures. Nowadays, CFD programs such as ANSYS® allow engineers to efficiently assess the response of offshore structures to various loads, whereas historically, this was often only possible through reduced-scale model testing in a physical ocean wave basin and through calculations on simplified approximate analyses based on first principles, which have limited accuracy. Moreover, the effort of producing a physical model and testing it means that physical testing is generally reserved until relatively late in the design phase; it is more suited for verifying an established design, rather than providing practical engineering data that could be used to drive the design process. In addition, any novel solution tested at model scale has increased uncertainty of actual performance at full scale due to delicate scaling issues. CFD programs are becoming increasingly sophisticated and they are capable of modelling non-linear interactions between both fixed and floating structures under a variety of wave types and sea spectra. These programs are especially useful for assessing complex structures such as jacket-type platforms, as shown in Figure 9.1.

While CFD programs are instrumental in the design process in terms of assessing the loads on a structure, it is not unusual for the loading to change over time. Having access to reliable, up-to-date loading estimates is important for several reasons. It informs engineers when assessing the eligibility

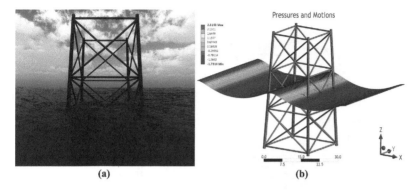

Figure 9.1 (a) Jacket-type structure and (b) wave loading assessment in a CFD environment.

of a structure for requalification schemes. The performance requirements on structures can change because of environmental reasons or because of changes that occur on the structure itself. Factors such as change in environmental exposure conditions or the installation of new, nearby structures can alter flow patterns and result in the actual operating conditions differing from what it was designed for. Additionally, changes to the structure itself, such as extensions, loss of carrying capacity in some members, and marine growth colonization can all affect the loading profile. Of these issues, marine growth colonization is particularly prevalent and hard to avoid. Some examples of marine growth colonization are shown in Figure 9.2.

The presence of marine growth on offshore structures increases the diameter and roughness of members. This creates several problems; most notably, it increases drag forces and creates unpredictable hydrodynamic instabilities (Ameryoun & Schoefs, 2013). These factors cause a loss in structural performance and reliability. Owners, therefore, have a keen interest in monitoring the progression of marine growth so that they can optimize cleaning regimes and have more reliable loading estimates. The diameter and roughness properties vary around and among structural components and require the full 3D shape of affected components to be

Figure 9.2 Examples of marine growth on structural members.

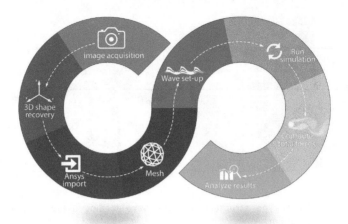

Figure 9.3 Workflow for building and analysing 3D finite element models from images.

computed. This is where stereo imaging can serve as a useful tool. The process for building and analyzing 3D finite element (FE) models from inspection imagery is presented in Figure 9.3.

This workflow is demonstrated on smooth and roughened cylinders that are fully submerged in a physical wave basin. The cylinders are 1 m in height and have a diameter of 140 mm. The roughened surface is created using ball-bearings that protrude from the surface of the cylinder by a maximum of 10 mm. The 1:25 scaling means that these cylinders correspond to monopiles having a diameter of 3.5 m at full-scale. The smooth cylinder represents a clean monopile without any marine growth, while the roughened cylinder represents a monopile that has been colonized by marine growth. The test cylinders are shown in Figure 9.4.

Step 1: The first step is to acquire the stereo images. This step involves encircling the cylinders with the stereo camera rig and photographing the cylinder from all sides, as depicted in Figure 9.5(a). The output from this step is a set of stereo image pairs.

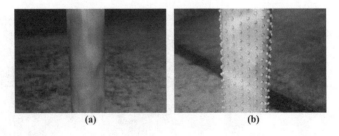

Figure 9.4 Test cylinders: (a) smooth—no marine growth, and (b) artificially roughened—simulated marine growth.

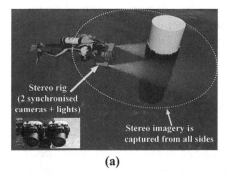

(a)   (b)

Figure 9.5 (a) Stereo image acquisition and (b) 3D shape profile of the roughened cylinder.

Step 2: The next step is the 3D shape recovery. Using the camera calibration and stereo matching techniques described in Chapter 7, the 3D shape of the cylinders is generated from the stereo images. The 3D shape profile of the roughened cylinder is shown in Figure 9.5(b). It may be noted that the surface perturbation details are apparent, which reflects well on the sensitivity of the 3D imaging process.

Step 3: The next step is to import the 3D data into ANSYS®. ANSYS® is a commercial software and is widely used as a standard tool for finite element analysis. The output from the 3D imaging process is a point cloud. In order to import this data into ANSYS®, it must be converted to a watertight, closed-volume surface. This is achieved by saving the 3D point cloud data as a PLY file. PLY is the Polygon File Format and was principally designed to store three-dimensional data from 3D scanners. The PLY file is opened in Meshlab, which is a free open-source program for the processing and editing of unstructured 3D meshes (Cignoni et al., 2008). From within Meshlab, Poisson surface reconstruction and other geometrical preparation operations can be performed such as healing and filling in gaps in the model (Campos et al., 2015). Once this is done, the mesh is exported as an STL file, which is a format that is commonly used by many CAD programs such as SolidWorks® and ANSYS®. The STL file is then ready to be imported into ANSYS®.

Step 4: The fourth step is to mesh the 3D model from within ANSYS®. Constructing a computational mesh is one of the most important parts in conducting a CFD simulation, and it always represents a compromise between accuracy and computational cost. ANSYS® provides several automated meshing tools, with adjustable parameters, which should be explored and fine-tuned by the user so as to find a mesh that works well for their model. A good mesh accounts for geometrical, physical, and numerical fidelity, as well as computational efficiency. The mesh may be "denser" in critical regions where the user would like to preserve finer details, while it can be courser at less important parts of the model. De-featuring can be employed here to remove excess detail not relevant to the simulation.

Step 5: Once the model has been meshed, the environment and wave conditions can be defined from within the Hydrodynamic Diffraction (AQWA®) analysis system. For this example, the focus will just be on regular waves of standard heights and frequencies/periods using Airy wave theory. Irregular waves, such as those formulated by the JONSWAP (Joint North Sea Wave Project) wave spectrum, can be studied as well. Additionally, current can be modelled to supplement the wave forces being applied to the structure.

Step 6: Once the wave and sea information have been set-up, a simulation can be run to obtain hydrodynamic and hydrostatic data, as well as pressures and motions information at various points on and around the cylinder, as illustrated in Figure 9.6. CFD simulations operate by subdividing the volume around the structure into a number of much smaller control volumes where the fundamental equations of fluid dynamics are applied.

Step 7: In order to compute total forces on the cylinder, a Hydrodynamic Time Response module needs to be added alongside the Hydrodynamic Diffraction module. The start time and end time of the wave-cylinder interaction are defined, along with the time interval that controls how often the force is computed. The simulation is run for the waves that were defined in step 5. The output from this step is a time series of wave loading.

Step 8: The final step in our process involves analyzing the results and comparing them to other sources. Historically, offshore engineers have often been unconvinced about the trustworthiness of results from CFD programs due to a number of challenges inherent in the process. However, advances in software and hardware technology have addressed many of those concerns, and now, modern CFD implementations are widely regarded as being a useful tool to augment, if not completely replace, physical testing in some cases. This section

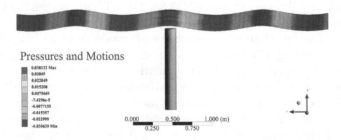

Figure 9.6 Pressure distribution around a cylinder.

looks at how the results obtained from CFD analysis compare against theoretical results, as computed using the semi-empirical Morison's equation, as well as results obtained experimentally through physical tank testing. Morison's equation states that the total force acting on a cylindrical body, $F_{Morison}$, is the summation of the drag and inertial force components, as per:

$$F_{Morison} = \frac{1}{2} C_d \rho D u |u| + C_m \rho \frac{\pi D^2}{4} \dot{u} \tag{9.1}$$

where C_d is the drag coefficient, u is the flow velocity, $\dot{u}$ is the acceleration of fluid, C_m is the coefficient of inertia, D is the cylinder diameter, and ρ is the mass density of the fluid. The force acts in-line with the wave direction.

The physical testing took place in a flume tank. It was identical to the CFD simulation in terms of the dimensions of the cylinders and the wave conditions. The wave loading was measured using load cells attached the smooth and roughened cylinders, as depicted in Figure 9.7.

Comparison of results reveals that the forces determined from each method (measured, theoretical, and from CFD simulation) agree very well with one other, as can be observed in Figure 9.8, which shows one example of wave loading on a smooth cylinder.

Validating workflows on simple geometries, such as these cylinders, gives engineers confidence that the approach is valid for more complex structures. Real-world inspections stand to gain from using 3D imaging techniques as it offers an efficient way of collecting in-situ measurements. However, it is important that the quantitative data obtained from these techniques can be effectively integrated into analysis software so that engineers can realize their full potential. This section describes and validates workflows for building FE models from underwater inspection imagery, and by doing so, demonstrates that quantitative imaging is a good starting point for undertaking more advanced engineering tasks.

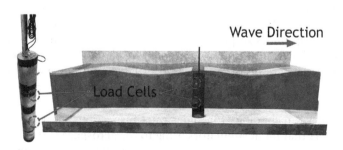

Figure 9.7 Flume tank experiment set-up: Load cells measure the wave loading on cylinders.

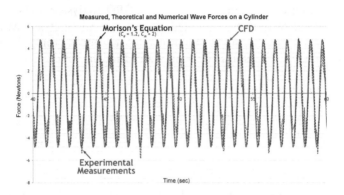

Figure 9.8 Measured, theoretical and numerical forces on a 140 mm diameter smooth cylinder that is subjected to waves of height 64 mm and period 1.4 seconds.

9.3 VIRTUAL REALITY INSPECTIONS AND SPHERICAL IMAGE ACQUISITION

Underwater inspection methodologies must continually be developed to meet the needs of fast-emerging technologies like floating wind turbines, wave devices, ocean-bed cables, risers, and long umbilicals. The capacity to test inspection methodologies and NDT tools without having to deploy full campaigns or large-scale experiments is of high practical value for inspectors. This section describes an approach for developing carefully controlled underwater virtual scenes that feature structural systems and using this as the basis for carrying out simulations that investigate methodologies of assessment in detail and in a realistic fashion, as well as for evaluating the performance of image-processing based NDT methods.

Virtual reality (VR) already has an established presence in other many other fields. A well-known implementation of VR technology is in the aerospace sector where flight simulators are routinely used for training pilots. These flight simulators can replicate a host of scenarios, such as flying in various weather and visibility conditions. They are useful for practicing and refining common tasks such as take-off and landing and coordinating with air traffic control, as well as for handling emergency events, where they help pilots respond to these situations in a safe environment. Additionally, VR is also quickly gaining a foothold in the consumer market. VR headsets are becoming more affordable and immersive, and there are also inexpensive and widely available VR glasses that are compatible with most smartphones.

Adopting VR technology for underwater inspections shares many of the same benefits that flight simulators bring to the aerospace sector. VR-based inspection simulations can help divers gain a better understanding of what to expect during inspections in a risk-free setting. Moreover, it can serve as an effective route-planning tool and it gives inspectors a chance to point out parts of a structure that require special attention, for example, critical joints, and easily relay this information to the divers. Communicating this information without the help of VR would otherwise be difficult and likely to introduce some level of uncertainty and confusion.

In this section, the process of developing realistic virtual underwater scenes is briefly explored. There are several software packages that can be used here, such as Blender (Roosendaal & Selleri, 2004), which is an open-source package that can be used for the creation, animation, and rendering of natural 3D environments. If the user has a 3D model of the structure saved in a recognized CAD format, such as an STL file format, then this may be imported into the Blender scene; otherwise, the structure can be built from scratch from within Blender.

An example of a virtual underwater scene is shown in Figure 9.9. It is displayed in a form that would be experienced as if viewing using a VR headset. This VR-ready image was created by spherically rendering the scene such that the whole environment around the camera is captured. This enables users to immersively look around in every direction when wearing the VR headset.

Another advantage of creating virtual scenes is that it offers a way of evaluating the performance of image-processing based algorithms. For instance, the performance of stereo matching algorithms for recovering the 3D shape of virtual marine growth colonized structures can be gauged since the depth/3D information is precisely known and available. This provides a unique way of imitating the texture and irregular shape of natural marine growth in an underwater setting while retaining complete control of the scene and the stereo cameras. It is also possible to adjust certain properties of the scene, such as lighting and turbidity levels, and investigate what impact this has on the performance of algorithms. An interesting point

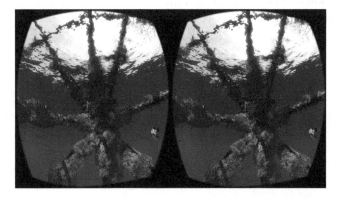

Figure 9.9 An example of an underwater scene developed in Vue. It is displayed in a form that would be experienced as if viewing using a VR headset.

here is that the synthetically created turbidity can be calibrated against the known and physically meaningful turbidity levels from the ULTIR repository (see Chapter 8) so that a strong basis in reality can be retained despite working in virtual environments.

9.3.1 Spherical image acquisition

Spherical cameras and virtual reality (VR) are closely related since spherical cameras are often used to obtain real-world VR content. In this sub-section, the role that these cameras can play in terms of improving the condition of monitoring is explored. Often with underwater inspections, the goal is to capture as much relevant information as possible within the short timeframe available to divers. This can be challenging for large or complex structures. Spherical cameras can help in this regard as they record in every direction at the same time. Capturing the whole environment around the camera allows inspectors to see more of the scene and it gives them the chance to look for instances of damage in all directions after the fact.

The two main types of systems for producing spherical imagery are smaller, cheaper spherical capture devices, multi-camera rigs comprised of several cameras pointing in different directions. The smaller, cheaper spherical capture devices tend to have small sensors and low image quality. This is exacerbated by the fact that the field of view encompasses the entire scene, which means that resolving fine details, such as cracks or small corrosion spots, in any one direction is difficult.

While multiple camera rigs are more expensive, the image quality is superior. Additionally, many of these rigs are aimed at the sports/outdoors enthusiast market and, consequently, they are rugged and waterproof. An example of a multi-camera rig is shown in Figure 9.10(a), which features 6 action cameras housed in a casing. It is a fully integrated solution with all cameras automatically synchronized. An example of a spherical image is shown in Figure 9.10(b). A drawback of creating spherical images/video with multiple cameras is the risk of artifacts appearing when the

(a)   (b)

Figure 9.10 (a) Multi-camera spherical imaging rig, and (b) example of a flattened spherical image.

constituent images are stitched together to form one complete spherical image. Effective camera calibration is crucial here for mitigating this problem. Moreover, the stitching process can be a bottleneck, especially if dealing with very high-resolution imagery.

One area where spherical cameras might find particular application lies in conducting cursory checks, such as searching for underwater debris pile-up around a bridge pier after a storm event, or as a precursor to more in-depth inspections where inspectors would like to get a rough sense of the extent of damage present prior to undertaking a more thorough assessment, as depicted in Figure 9.11. In these cases, a spherical camera may provide the necessary information and offset the need to hire a diver. Adopting new

Figure 9.11 Spherical may find applications for frequent, low-level inspection tasks.

acquisition methodologies is only one ingredient in the broader evolution of image-processing in underwater inspections. Algorithmic advances will also help to shape the future of this field.

9.4 DEEP LEARNING BASED DAMAGE DETECTION

Given the value that automatic damage detection techniques can bring to the inspection process in terms of increased efficiency and convenience, it is worthwhile developing and investing in new high-performing methods. Deep learning techniques have attracted significant interest in recent times as they have produced impressive results on benchmark and real-world data sets across a wide range of computer vision tasks. They offer numerous advantages over conventional techniques—they can learn rich feature representations that are often more robust than traditional hand-crafted features. Moreover, deep learning models are highly repurposable. For instance, a corrosion detection model that is trained on a large dataset of corrosion images can be reused as a crack detection model by simply training the model on crack images instead. Very minor changes to the structure of the model, if any, are necessary. In contrast, a conventional corrosion detection technique cannot be easily reformulated to function as a crack detection technique, as typically, each would follow completely different methodologies. The main obstacles that have stifled deep learning techniques in the past are the lack of large training datasets and the need for efficient GPU-accelerated computing. However, the increasing availability of labelled images featuring various damages, along with modern hardware advances, have now made deep learning widely accessible. This section suggests ways to get more mileage from existing damage datasets, even if these datasets only relate to above-water instances of damage. Moreover, many pre-trained models now exist and can be used to bootstrap custom models with relatively small datasets (i.e., around 100 images).

Leveraging an existing network that has been pre-trained on a large dataset is a good starting point for building customized damage detection techniques. The weights of the top layers of the existing network can be fine-tuned using only the limited data that is available. Such an approach allows a high degree of accuracy to be attained in an efficient manner. Furthermore, users can extract more value from the limited training data by carrying out a number of random geometric transformations and optical adjustments so that the model would never see twice the exact same picture. This helps prevent overfitting and allows the model to generalize better. Performing color adjustments to the training data so as to emulate the underwater environment should help to develop models that are better suited for classifying imagery obtained from real underwater inspections.

As an example, this section demonstrates an image-processing method that uses a convolutional neural network (CNN) architecture to detect instances of corrosion. CNNs are a pillar algorithm in the domain of deep learning. The multi-layered network was trained using a dataset of 120 still images taken from a purpose-built repository that is available at http://www.ultir.net/. These 120 images were extended to 600 images using the process depicted in Figure 9.12, that is, four new images were generated from each original image using a variety of random transformations.

Damage detection could be performed by scanning through the entire image using a sliding window based approach and classifying each window as representing damage or non-damage. However, this is a highly computationally demanding approach, especially if dealing with high-resolution imagery or video footage where many frames must be analyzed. A more efficient approach involves using k-means clustering to segment the image into many regions, erring on the side of over-segmentation. Each region is then classified using the CNN. This approach enables near real-time computational performance, which makes it reasonable to efficiently sift through hours of inspection footage, or archival footage from past inspections, and extract instances of damage for further scrutiny. Sample results using this approach are presented in Figure 9.13. These results relate to inspection footage where corrosion was present on a submerged pile and a seawall.

The positive results support the notion that data-based image-processing methods such as CNNs offer many benefits over traditional model-based methods, and that such methods can serve as a useful addition to the inspector's toolbox, increasing reliability and repeatability of inspection findings.

Figure 9.12 Process for generating more training samples.

Figure 9.13 Sample detection of a deep learning based corrosion detection technique applied to inspection video footage.

9.5 VIDEO ANALYSIS

So far in this book, the focus has largely been on still images and sequences of images without any time component. Video analysis could be considered to be an extension of image analysis since, in addition to pictorial information, video analysis also has access to temporal information that captures scene evolution. Video analysis has the potential to be used as a substitute for many instrumentation devices that are currently being used to monitor the health of marine structures. This section looks at one such application as a concrete case study. It involves characterizing the dynamic response of a pedestrian bridge using video tracking. The dynamic nature of a bridge provides an indication of its serviceability condition and the level of dynamic acceleration can be related to the comfort level experienced by bridge users (Van Nimmen et al., 2014). Additionally, a knowledge of the bridge parameters, such as the natural frequency and mode shapes, is useful for numerous applications, including for verifying or strengthening finite element (FE) models so that they more accurately reflect the true behavior of the structure. This aspect has attracted a growing interest in recent years as FE modelling is increasingly being utilized as a powerful and effective assessment tool.

Vibration-based assessments are carried out using either direct or indirect instrumentation techniques. Monitoring via direct instrumentation usually requires the bridge to be outfitted with a range of sensors placed directly on the bridge, which is generally expensive and time-consuming. These sensors must be rugged enough to cope with the marine conditions. The development of wireless sensor networks in recent times has partially alleviated this issue by adding a degree flexibility and value compared to permanently wired systems. Indirect vibration based assessments typically involve vehicles equipped with onboard sensors traversing a bridge and recording the induced vibrations. Such an approach is attractive in cases where the cost associated with permanent instrumentation and routine inspections cannot be justified.

Video analysis is an alternative to these conventional approaches. It is suitable for analyzing bridge vibrations that are characterized by large amplitudes and low frequencies. Video tracking has the advantage of being a low-cost and easily accessible option as the only hardware required is a standard digital camera capable of recording video. The onsite set-up and video acquisition is straightforward and does not require lengthy configuration or calibration procedures. Furthermore, the acquired data is easy to interpret as the excitation sources can be identified in the video itself. The illustrated video tracking approach involves focusing a camera at the midspan of a bridge and tracking the precise location of a point on the bridge in each frame of the video as it is undergoing user-induced vibrations. The time series data related to the location of this tracked point is then analyzed to obtain the frequency domain response.

Figure 9.14 Daly's Suspension Bridge in Cork, Ireland, and the location of a mounted video camera on the river bank.

The approach is demonstrated as a proof-of-concept on a historic steel pedestrian suspension bridge in Ireland, called Daly's Bridge, which is known locally as the "Shaky Bridge" due to its notable and, for some, unnerving swaying motion under pedestrian loading. The bridge is shown in Figure 9.14. This bridge is an ideal candidate for video-based analysis as the dynamic displacement responses of the bridge are large enough to be detected in the video. The frequencies obtained from video analysis are checked against accelerometer data obtained from sensors on the bridge.

9.5.1 Equipment and set-up

There are a number of equipment-related factors that have a key impact on the performance of video tracking based vibrations assessments. First, the maximum frequency of vibration that can be encoded is controlled by the sampling rate, or the video frame rate. As an example, the digital video camera used in this case study was a Canon 600D, which recorded video at a rate of 35 FPS with a higher resolution of 1920 pixels × 1080 pixels. Second, the minimum amplitude of the vibration that can be detected in the video is largely affected by the optical power of the lens. Long-focus lenses are able to make faraway objects appear closer. This effect means that a long-focus lens would magnify bridge displacements and make them more apparent in the acquired video. A 55 mm equivalent lens (which was closer to 90 mm when accounting for the crop factor—see Chapter 3 for an explanation on this) was used in this case study. This focal length was sufficient as, even though the camera and the target on the bridge were sperated by quite an extensive distance, the bridge oscillated at large amplitudes.

9.5.2 Video tracking technique

The video tracking technique operates by tracking the movement of a point on the bridge while in an excited state. Tracking is done by picking a small patch, or window, which is centered on the chosen point in the first frame of a video sequence and following it throughout the duration of the video clip. For every successive frame, the point is located by finding the patch that best correlates with the patch in the previous frame. The best correlation can be found using some measure of similarity. Common measures of similarity include sum of squared distances (SSD), sum of absolute distances (SAD), normalized cross-correlation, and zero-mean normalized cross correlation (ZNCC). Similarity measures such as SSD and SAD are highly efficient; however, this comes at the expense of reduced tracking accuracy and increased sensitivity to noise.

For video tracking, the window centered on the tracked point in a frame is matched to the corresponding window in the next frame. The search space for locating the corresponding window is confined to a stationary region in the video, which must be carefully chosen such that it encloses the tracked point throughout the entire video. Confining the search space to the local neighborhood allows for greater computational parsimony and minimizes the risk of false matches.

For this example, the ZNCC is used as a matching metric since it is widely regarded as being a robust and reliable measure of similarity. The ZNCC score has a high value when there is a high degree of similarity between a patch in the reference frame and the succeeding frame. The point in the succeeding frame corresponding with the highest ZNCC score is taken as the new location of the tracked point. This procedure is repeated until the final frame in the video sequence. The pixel locations for the tracked point are recorded at each frame, which may be viewed as a time series data of dynamic displacement responses. Spectral analysis of this time series is carried out to identify the major frequency components.

9.5.3 Tracking challenges

There are a number of challenges associated with tracking a moving point in a video sequence. Most notably, the tracked point may drift or become completely lost as a result of temporary occlusion or luminous complexities in the scene such as shadows or light reflections. To mitigate these issues, a prominent and distinct point (i.e., a patch in the image with high local contrast) should be selected for tracking. If needed, multiple trials can be carried out by selecting various high-contrast regions near the mid-span of the bridge and assessing how well they are tracked over the duration of the video.

Another challenge relates to locating the position of the tracked region as precisely as possible. Even at the highest video resolution, the motion of

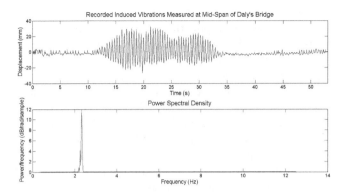

Figure 9.15 Vertical displacement versus time graph measured at mid-span of the bridge, and Fourier spectrum of vibration data.

the bridge translates to minute changes of only a few pixels in the video frames. In order to address this, sub-pixel and sub-scale interpolation can be employed as described by Brown and Lowe (2002).

9.5.4 Results

The displacement versus time graph corresponding to a pedestrian-induced vibration scenario is shown in Figure 9.15, along with the power spectral density (PSD) plot, which is found by taking the Fourier spectrum of this dynamic displacement response.

In Figure 9.15, a prominent peak may be observed at the estimated natural frequency of 2 Hz in the power spectral density plot. Accelerometers placed at the mid-span of the bridge indicate similar values for the bridge's natural frequency. These results validate the proposed video tracking method as a convenient way of identifying the natural frequency of bridges characterized by large amplitudes and low frequencies to a reasonable degree of accuracy.

9.5.5 Smartphones

In this section, we will consider the merits of using smartphone cameras in lieu of conventional cameras. The ever-improving image quality of smartphone cameras and their expanding feature-set may mean that these devices can play an increased role in future inspections. In this section, their qualities are analyzed and they are compared with dedicated imaging devices.

Much like the previous example in which video analysis was used to characterize the dynamic nature of a bridge, in this exercise, the potential of smartphones to act as an adequate alternative to high-speed cameras for detecting transient vibrations is investigated. For most affordable digital cameras, the shooting rate is typically limited to 30–60 frames per second (FPS). For some engineering applications, this introduces the need for high-speed cameras. By using high-speed cameras, the flexibility and the range of applications for which image-processing can be used increases greatly. Normally, high-speed cameras are costly and difficult to use due to the amount of memory required due to the high number of frames and the computational resources needed to effectively process the video. Smartphones have now created a possibility of taking video images at higher speed, but are not well compared against established benchmarks. For the purposes of this exercise, a Phantom® v5.1 high-speed camera (Figure 9.16(a)) and the slow-motion function of a Nexus® 6P smartphone (Figure 9.16(b)) were used to capture the motion of an oscillating system.

The Phantom® v5.1 can shoot video up to 1200 FPS at a resolution of 1024 × 1024 pixels. However, this high sampling rate is largely redundant for characterizing the dynamic nature of structures/structural components, where sampling rates in the order of 200 FPS are often sufficient. A sample frame captured by the Phantom® v5.1 high-speed camera is shown in Figure 9.17(a). In comparison, the slow-motion function of Nexus® 6P smartphone can record color video up to 240 FPS at a resolution of 1280 × 720 pixels. A sample frame for this camera is shown in Figure 9.17(b).

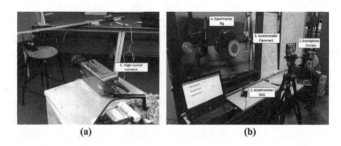

Figure 9.16 Video acquisition using (a) a Phantom® v5.1 high-speed camera and (b) a Nexus® 6P smartphone.

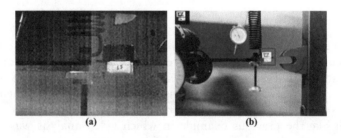

Figure 9.17 (a) Sample frame captured by the Phantom® v5.1 high-speed camera at 200 FPS, and (b) sample frame captured by the Nexus® 6P smartphone at 240 FPS.

The target of interest is indicated by the red bounding box in Figure 9.17(b). The position of this target is tracked over the duration of the video to determine the vibration frequency.

Visually, it may be seen that the Nexus® 6P smartphone boasts superior image quality over the Phantom® v5.1. The image quality has a major impact on the ability of image-processing based tracking algorithms to successfully track an object on a vibrating structure, and it is thus important to maximize the quality of the input footage. In this example, the tracking algorithm was ultimately able to track the point of interest for both the smartphone and the Phantom® v5.1 videos, despite the reduced quality of the later. Such an outcome supports the idea that smartphones can be a reliable and effective option in certain cases. Moreover, this case study revealed some clear advantages of using smartphones over conventional high-speed cameras from a logistical and operational standpoint. Beyond their low cost and ubiquitous nature, smartphones have the advantage of being considerably easier to set up and use, the output video file size was more manageable (the output video size was ~20 megabytes versus several gigabytes for the Phantom® v5.1), and a range of applications, or "apps," could easily be installed that extend imaging functionality.

In summary, the results of these tests show that the video-based motion tracking technique is a viable, low-cost method for executing quick vibration based assessments of systems. The evaluation of smartphone cameras also addresses the ongoing uncertainty around the level of performance that can be achieved using cheaper sensors for quantitative purposes. In the illustrate case study, it was found that smartphones could act as practicable video acquisition devices. Such findings provide reassurance for inspectors wishing to use their smartphones for simple monitoring tasks.

9.6 USE OF EXISTING IMAGE ARCHIVES FROM PAST INSPECTIONS

Damage from past inspections is usually photographed and archived by the inspector. These historic photographs can reveal valuable information about the manner in which a structure is deteriorating and the rate at which this is occurring. In the same vein, photographs captured as part of present-day inspections will be a useful reference for future inspections. However, a challenge here is that the in-service life of marine structures spans several decades. During this time, improvements to camera technology and new image acquisition practices mean that the imagery captured from past inspections may be somewhat deficient compared with current imagery, and thus, comparisons cannot be easily drawn, at least on a like-for-like basis and in a quantitative sense.

Nevertheless, there are ways in which quantitative information can be extracted from archived inspection imagery if we allow ourselves to make some assumptions about the scenes based on monocular cues as well as our physical understanding of objects in the scene. With reference to Figure 9.18(a), the only piece of information we know is the width of the prong at the end of the rod, and we wish to determine the diameter of the mooring line that has become covered with mussels. Because the end of the rod appears to be near the mooring line, it can be quite safely assumed that the physical sizes of pixels at the end of the rod and on the mooring line (at the point closest to the rod) are similar. The diameter of the mooring line at this point can be measured in pixel units and then converted to physical units (e.g., centimeters). It may be helpful to construct a 3D model that follows the outline of the mussel-covered mooring line. Once the 3D model is created, it becomes easier to read off dimensions.

Another way to extract information from archived inspection imagery is by overlaying images captured at different inspection campaigns. An example of this is shown in Figures 9.18(b) and 9.18(c). Figure 9.18(b) represents moderate marine growth colonization, while Figure 9.18(c) represents more extensive marine growth colonization, as might be expected after a few more years. It is not essential that the photographs are captured from the exact same position and with the same focal length; however, less variation

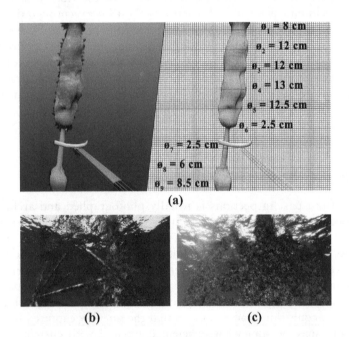

(a)

(b)   (c)

Figure 9.18 (a) Extracting physical information by making an assumption in the scene. Comparing photographs taken as part of different inspection campaigns to reveal changes/progression of damage: (b) earlier inspection—moderate marine growth colonization, and (c) later inspection—more extensive marine growth colonization.

between photographs makes it easier to draw meaningful comparisons. One image can be transformed (moved, rotated, perspective transformed) with respect to the other image so that points of interest in the images overlap as much as possible. Using this approach, changes in the structure should be easily identifiable.

Going forward, provisions should be made to future-proof the imagery as best as possible. This includes following a sensible naming convention that encodes contextual information such as where the damage is on a structure. Additionally, the date and time should be correctly set in the camera(s) as this information will be embedded in the EXIF (exchangeable image file format) data for each photograph. On this note, care should be taken that the EXIF data is not lost if saving the photographs in a new file format.

9.7 SUMMARY

A key objective of this chapter is to show that image-processing in underwater inspections is advancing on many fronts. Ways in which image data can be exploited to a greater extent by integrating it with specialist engineering software for carrying out high-level analysis are presented. New applications for emerging technologies such as virtual reality (VR) and spherical imaging are discussed, and we examine how these can play a role in the inspection process, both in terms of improving the condition of monitoring and obtaining more information from underwater inspections, as well as using this technology to plan routes and brief divers. As part of this, the motivations and tools for developing virtual 3D models of structures and visualizing them in realistic environments are explored. This chapter also looks at algorithmic advances, specifically in relation to deep learning techniques. Such techniques have attracted significant interest in recent times and are key to image-processing. The benefits of these techniques are discussed, such as their highly transposable nature, whereby a corrosion detection model can have the same framework as a crack detection model. Additionally, the increasing availability of labelled images of various damage forms bodes well for the future of these techniques. Ways to get more mileage from existing damage datasets are suggested.

Video analysis is investigated as this has the potential to be a powerful tool for many applications in engineering and Structural Health Monitoring (SHM), including for the assessment of marine structures. We focus on one such application in which a video tracking based approach was used to characterize the dynamic response of a pedestrian bridge that spans a river. On top of this, we look at how inspectors can capitalize on the ubiquity of smartphones. It was found that they can act as viable video acquisition devices for quantitative imaging purposes. The positive results obtained using smartphones lends support to inspectors wishing to use their smartphone devices for simple monitoring tasks in the future.

REFERENCES

Ameryoun, Hamed, and Franck Schoefs. "Probabilistic modeling of roughness effects caused by bio-colonization on hydrodynamic coefficients: A sensitivity study for jacket-platforms in Gulf of Guinea." In *ASME 2013 32nd International Conference on Ocean, Offshore and Arctic Engineering*, pp. V001T01A057–V001T01A057. American Society of Mechanical Engineers, 2013.

Brown, M., and D. Lowe. "Invariant features from interest point groups." Proc., BMVC2002: British Machine Vision Conference 2002, 2–5 Sept., British Machine Vision Assoc., pp. 253–262, 2002.

Campos, Ricard, Rafael Garcia, Pierre Alliez, and Mariette Yvinec. "A surface reconstruction method for in-detail underwater 3D optical mapping." *The International Journal of Robotics Research* 34, no. 1 (2015): 64–89.

Cignoni, Paolo, Marco Callieri, Massimiliano Corsini, Matteo Dellepiane, Fabio Ganovelli, and Guido Ranzuglia. "Meshlab: An open-source mesh processing tool." In *Eurographics Italian Chapter Conference*, vol. 2008, pp. 129–136. 2008.

Roosendaal, Ton, and Stefano Selleri, eds. *The Official Blender 2.3 guide: Free 3D creation suite for modeling, animation, and rendering.* Vol. 3. San Francisco: No Starch Press, 2004.

Van Nimmen, Katrien, Geert Lombaert, Ilse Jonkers, Guido De Roeck, and Peter Van den Broeck. "Characterisation of walking loads by 3D inertial motion tracking." *Journal of Sound and Vibration* 333, no. 20 (2014): 5212–5226.

Chapter 10

Conclusions

10.1 SUMMARY

This book provides a guide to image-processing techniques for engineers and inspectors who wish to harness the full potential of cameras as an inspection tool for built infrastructure. It is principally geared toward the inspection of the submerged part of offshore and marine structures as there is a paucity of research material related to the use of image-based methods in underwater inspections, despite being well-suited to the task. The general goals of this book pivot around image-processing methods for Structural Health Monitoring (SHM), assessment and inspection related applications, and showing the benefits that adoption of such methods can have for the management of offshore and marine structures.

Traditionally, image collection for built infrastructure inspection and assessment was often carried out on an ad hoc basis, and as a result, attempts to extract meaningful quantitative information were built on weak foundations. A key objective of this book is to outline and nurture good image acquisition practices so that high-quality imagery can be obtained on a consistent and reliable basis. We set about this by describing a protocol for image collection, detailing camera set-up and on-site calibration procedures, and providing advice on how to handle the underwater visibility conditions. These guidelines prepare the acquired imagery better for application of image-processing algorithms.

This book covers the fundamentals of image-processing and subsequently presents several image-processing techniques related to damage assessment. These techniques are demonstrated on a number of real-world applications. The techniques engage with features of interest with different geoemtric dimensions in the form of a crack detection technique, two surface damage detection techniques (one based on color information and one based on texture information), and a 3D imaging technique using a system based on stereoscopic imagery. The crack detection and surface damage detection algorithms are the building blocks for most damage detection tasks, while 3D imaging is useful as it can be used for volumetric measurements, documentation, and general visualization. This book provides a step-by-step breakdown of how the chosen solutions are designed and deployed, along with practical insight into why certain design choices are made. Readers can run these techniques themselves using the MATLAB® scripts included with this book. Moreover, access to an underwater image repository means that readers can adapt the algorithms to their specific needs and conveniently evaluate the performance of their algorithms.

The principal benefits of image-processing stem from the fact that it adds an objective and quantitative source of information to complement the largely qualitative information obtained from visual inspections. The quantitative nature of the data naturally lends itself to be integrated into the decision-making process and to be used in an endless array of upstream analysis activities. In this book, we try to adopt a more exploratory approach to give readers a feel of what kind of analysis activities can be undertaken. We pay special attention to high-value use cases for engineers/inspectors such as using imaging data as input for finite element analysis and computational fluid dynamic simulations. Establishing these workflows empowers engineers/inspectors by enabling them to quickly and accurately perform analysis and thereby allowing them to make better and faster decisions. The authors hope that by showcasing these examples and workflows, it will encourage readers to think about the role that imaging can play as part of their projects, as a key aspect of harnessing the power of image-processing is the ability to identify which tasks are suited to imaging and what factors can impede performance.

10.2 LIMITATIONS AND FUTURE RESEARCH DIRECTIONS

Image-based methods undoubtedly have value as a quantitative inspection tool. However, let us briefly consider some of the inherent limitations. First, the technology is only appropriate for assessing visible damage forms and surface breaking defects. It is not possible to assess internal defects or damage that is masked by bio-fouling without first cleaning the structure. Second, the performance of image-based methods is often hampered by the poor visibility and complex underwater light field. This book sheds some light on this problem by presenting a repository driven approach for evaluating the performance of methods under various visibility conditions. Finally, the ability of image-based methods to detect damage is heavily reliant on the vision of the operator to spot instances of damage and photograph them. Speaking in more general terms, there is still a wide gap between image processing algorithms and humans when it comes to semantic scene interpretation. Image processing techniques can struggle with tasks that are intuitive to humans. As such, we may not yet be at a state where inspectors can fully rely on image-processing techniques without some level of supervision or oversight. However, the field of computer vision is continually advancing and emerging algorithms show promising signs that should help bridge this gap. We are already seeing an increase of new and improving image acquisition devices and technologies that will help shape the future of image-based underwater inspections.

As of now, image-based underwater inspections is not a very developed field of study, and there is scope to advance the field in several directions. In the near future, the authors believe the following research areas will attract increased attention:

- Development of smart ROV based inspections, where onboard light meters and turbidimeter sensors automatically influence optimum distance between the ROV and the structure and automatically adjust the intensity of the artificial light sources, so as to maximize the quality of the inspection imagery and consequently maximize the detection of damage. A computerized database of illumination levels and turbidity levels can be the basis for choosing the optimum distance.
- As multi-camera systems become more popular, there will be a growing need to efficiently sift through the huge amounts of data that these systems produce and rank instances of damage that should be prioritized for the attention of inspectors. This is also highly applicable for autonomous inspections where the onboard video cameras are always recording and much of this data will be useless. This task will require efficient algorithms.
- On a related note, the increasing amount of labelled visual data on the Internet, even datasets that are specifically related to underwater damage that have been created on the back of this book, should pave the way for the development of more performant data-driven algorithms. The power of imaging in underwater inspections will be greatly extended once algorithms have matured to a level where they can be applied to almost arbitrary collections of inspection imagery and produce promising results.

The breadth of the challenges inherent in this field, combined with the richness and the utility of the image-processing techniques, will ensure that this remains an exciting field of study for many years to come.

As a final takeaway, this book can be considered to be an initial course for engineers and inspectors on how to effectively integrate imaging into their inspection regimes. If we are ever to realize the full potential of this technology, we must have many more properly trained personnel who can contribute to this topic through applications or advance the detection techniques. It is hoped that this book can help in this regard. The rewards, in terms of improved methods for monitoring our marine infrastructure, have noteworthy operational and analytical implications.

Index